JUTE CULTIVATION IN INDIA – THEORY AND PRACTICES

Prof. M.M. ADHIKARY
Former Dean, Faculty of Agriculture
BIDHAN CHANDRA KRISHI VISWAVIDYALAYA
Mohanpur – 741252, Nadia, W.B.

Dr. D. SAHA
Reader, Deptt. of Ag. Chem. and Soil Science
BIDHAN CHANDRA KRISHI VISWAVIDYALAYA
Mohanpur – 741252, Nadia, W.B.

Dr. S.K. BISWAS
Professor, Directorate of Jute Development
Ministry of Agriculture, Govt. of India
Nizam Palace, Kolkata, W.B.

KALYANI PUBLISHERS
LUDHIANA - NEW DELHI - NOIDA (U.P.)
HYDERABAD - CHENNAI - KOLKATA - CUTTACK

KALYANI PUBLISHERS

Head Office :
BI-1292, Rajinder Nagar, **Ludhiana**-141 008

Administrative Office :
4779, Ansari Road,
24, Daryaganj, **New Delhi**- 110 002

Works :
B-16, Sector 8, **NOIDA** (U.P.)

Branches :
1, Mahalakshmi Street,
T. Nagar, **Chennai**-600 017

110/111, Bharatia Towers Badambadi
Cuttack-753009 (Orissa)

3-5-1108, Narayanaguda, **Hyderabad**-500 029 (A.P.)

10/2B, Ramanath Mazumdar Street,
Kolkata-700 009

B116105061

Typesetting by :
(JUTE)
Pyramid Enterprises
215-Sangam Complex, Milap Chowk,
Jalandhar-144001. Ph.: 0181-2213422

81-272-2457-X

PRINTED IN INDIA

At Kalyani Printings B-15, Sector 8, NOIDA (U.P.) and Published by Mrs. Usha Raj Kumar for Kalyani Publishers, New Delhi-110 002

PREFACE

Jute and mesta are the two most important crops in the country particularly for eastern and north-eastern states. Raw jute comprising jute and mesta, provides sustenance to about four million farm families and about 2.5 lakh industrial workers. Besides, about 25 lakh people are connected in jute based ancillary sector. It has got high socio-economic importance and it is considered an eco-friendly crop. In spite of having high importance of the crop, it has to face, lot of competition with synthetic product as well as with other jute growing countries in the world market. In recent years jute is gaining lot of importance due to the new dimension for its use in the manufacture of diversified products and its environment friendly nature. For sustaining the jute sector, it is necessary to increase the productivity and to improve the quality of fibre. Govt. of India has taken special interest in this regard.

Jute is a fibre crop and its production depends upon efficient agronomic as well as cultural practices. Apart from that post-harvest operations of jute are important factors for getting good quality fibre. Marketing of jute, diversified use of jute fibres as well as quality jute seed production are to be dealt seriously for maintaining the quality characters of jute in future days. In this book eleven chapters have been included. All these chapters discuss different aspects of jute starting from origin and history to cultural management as well as post-harvest operations, use of raw jute and seed production technology for getting good quality plants for future generations. The authors have no hesitation to say that helps of different lectures delivered in different workshops and seminars by renowned scientists related directly or indirectly to jute cultivation have been considered during the preparation of this book. It is true that all aspects of jute cultivation have not been covered up in this book but an effort has been made to concise all the major consensus package of practices related to jute cultivation in a single volume.

Constant eye during preparation of manuscript can not even attain some general mistakes. Any kind of healthy criticism to improve the quality of book will be attended with great sincerity and appreciation.

M.M. Adhikary
D. Saha
S.K. Biswas

CONTENTS

CONTENTS

INTRODUCTION

Jute and mesta are fibre yielding crops and rank second in importance only after cotton in this category. In trade and industry both the crops together is known as raw jute. The fibre is obtained from stem (bark) and is known as bast fibre or phloem fibre. The jute fibre is obtained from two species belonging to *Tiliaceae* family viz. *Corchorus capsularis* known as white jute and *Corchorus olitorius* known as tossa jute. While mesta fibre is obtained from two species belonging to *Malvaceae* family viz. *Hibiscus cannabinus* and *Hibiscus subdariffa* var. *altissima* known as mesta or bimli in India and as kenef in other producing countries. However, *H. subdariffa* is also known as roselle. The cultivation of jute is confined mainly to the eastern region of the country covering seven states namely, West Bengal, Bihar, Assam, Orissa, Uttar Pradesh, Tripura, Meghalaya. West Bengal alone accounts for about three fourth of the total production and Assam and Bihar may be named as other two states giving appreciable quantity of production. Again, in the eastern states although eighty five districts have been identified as jute growing districts only twelve of these are major production districts. Hence, this crop bears a great deal of physiographic and socio-economic selectivity.

Jute is a crop with bright past, cloudy present and prospective future. During past days jute occupied a pride place amongst the crops. In terms of face value, one md. of jute was equivalent to three mds. of rice. It was genuinely called as the "Golden fibre of Bengal". The bright lustrous fibre was the epitome of fortune and happiness of farmers.

During the past two hundred years, or so, jute has made so much in road into the social structure of Bengal that even the crop has lost much credibility about its economic significance but its social value is still enormous. During these days of ever growing unemployment and under employment, jute being a labour intensive crop, a crop area of 0.86 million ha can generate a man-days of 215 million in rural areas for production of the crop. In the industrial front it provides direct employment to 2,50,000 workers and in ancillary activities like, transportation and trading it can provide seasonal occupation to 2.5

million people. Sale of raw jute in a short span is a sort of festival bonus to 4 million farm families during pre-festive seasons of financial crunch.

In the jute growing areas jute sticks are the main source of fuel. In view of the shortage of wood and coal, jute cultivation occupies an important position as a source of fuel in the rural areas. After harvesting the jute crops, the leaves are allowed to shed in the fields, which are decomposed in the soil and served as a source of manure resulting in the increase in soil fertility. In the jute field, the jute follows rice in many of the areas. At the time of harvesting the jute field remains almost free of weeds and resulting less weed competition in the following crops.

Apart from the above, jute is a natural and renewable source of raw materials and biodegradable as a result of which jute is considered as environment friendly crop. The proliferation of jute based economy on rural families is so elaborate and total that despite steady rise in cost of production and almost near stagnant market price, neither the policy makers nor the farmers have dared to impose a total ban on its cultivation. But jute is definitely slipped down from its "golden days of golden fibre" and has now miserably stabbed with a brand of non-remunerative crop.

The following salient information should be taken into consideration before jute cultivation.

- Yield per unit area is more with 'tossa' varieties than with white varieties of jute.
- Total replacement of 'white' jute by 'tossa' even if possible, is not desirable.
- The jute industry has not indicated its preferences; it uses both tossa and white jute.
- Interspecific hybridization between tossa and white has not given any hybrid.
- Pattern of rainfall largely determines which of the two species is to be grown.
- Late February to mid-March sowing should be with white jute varieties.

- Jute sowing starts in late February in soils retaining moisture of previous flood and harvested in June.
- Jute sowing is marked by norwester showers followed by a months lull (absence of rain) and subsequent monsoon showers.
- Alternate sunshine and rain and drizzle are conducive to good growth.
- Optimum rainfall is between 1500 and 2000 mm.
- The optimum spacing is 20 cm between rows (tossa) and 30 cm (white) ; high fertility does not support larger spacing within the rows i.e. 30 cm average.
- Jute grows best on soils where silt deposition due to river flood is an annual feature.
- Loamy soil is better than clay-loam; heavy clay is not suitable.
- Highly acid soils (below pH 6.0) is not conducive to good growth.
- Water logging before thrity days crop age is detrimental for both tossa and white jute.
- White jute can tolerate water logging at later stages of growth but tossa jute cannot withstand waterlogging.
- Water logging for prolonged period forces growth of adventitious roots resulting in 'mossy' fibre.
- Land preparation and sowings are best done with norwester showers; the period of lull that follows facilitate weeding and thinning and monsoon ensures growth.
- Premature flowering is not linked with age of seed.
- Badly stored old seed that has started losing viability may flower prematurely if exposed to drought.
- All varieties of jute may flower prematurely with prolonged exposure to low humidity (19% and below) and high temperature (37°C).
- JRO 632, JR-66 may not be sown earlier than mid April to avoid premature flowering.
- JRO 524, JRO 7835, JRO 878, JRO-8432, JRO-128 can be sown from 23rd March onwards.
- Uniform placement of seed (for both broadcasting and row cropping) demand a well pulverised weed free soil.

- Weed clearing during soil preparation reduces cost on subsequent weeding in post emergence phase.
- Jute responds readily and positively to proper thinning and application of nitrogen fertilizer.
- Nitrogen can be fed as urea through foliar spray.
- Early harvest (less than 110 days) ensures good quality but with sacrifice of yield. Late harvest gives high yield but produces coarser fibre with more of the unretted bottom called roots.
- Currently premium for quality does not cover loss in yield due to early harvest.
- Some amount of cowdung manure should be used along with inorganic fertilizers.
- Almost sixty percent of jute farmers grow jute in small plots (1 bigha or less) depending on family labour.
- Jute farmers who hire labour find jute growing unattractive.
- Row cropping with drills is possible in upland loam but maintenance of drills is doldrums.
- The parts of drills marketed by several manufacturers are not always interchangeable.
- Improved technique calls for larger investment. The raw jute price is not always that attractive to enthuse farmers, specially so as prices are not steady.
- More area can be brought under jute in jute-paddy double cropping system.
- Facilities for proper retting is shrinking fast.
- Retting is the most important factor that makes the quality.
- Technology for machine decortications and enzyme retting is coming up; there are no takers.
- Mills may each adapt a village to boost yield and quality.

CHAPTER 1

ORIGIN AND DISTRIBUTION

Genus Corchorus, to which jute has 40 species, but only two are cultivated; the remaining are all wild. Of these 40 species, greatest diversity is found in East African region, but those are all Chorchorus olitorius type. Thus, it is considered that East-African region is the area of origin for olitorius.

Survey work of other species has revealed that Indo-Burma border area containing many wild species of Capsularis, may be the area of origin of the species. There is another school of thought, which considers Southern China as the possible alternative area of origin of Capsularis.

From the point of view of distribution, jute cannot be branded as International or National crop as its cultivation is very much confined to some specific regions of some countries. Thus, in the International front India and Bangladesh account for 75 p.c. of the total production. The other jute producing countries are China, Thailand, Brazil, Peru, Burma, Nepal and Vietnam.

In India, as has been mentioned earlier, its cultivation is localised in Eastern states, West Bengal, Bihar and Assam being the major share holders. Out of total area of 0.86 million hectare, the share in area of leading states being about 0.63, 0.14 and 0.08 m.ha respectively.

In West Bengal again the same trend of localised cultivation is evident. The average area (9th plan) in the leading districts may be approximately as follows :

	in '000 ha
Murshidabad	140
Nadia	132
Coochbehar	85
North Dinajpur	60
24 parganas(N)	57
Jalpaiguri	47
Hooghly	32
Malda	25
Burdwan	15

Hence even in West Bengal which is a premier state in jute production, there is a clear indication of concentration of production and only five districts can be termed as leading districts. However, year after year there is a variation in coverage of areas in different districts.

CHAPTER 2

CLIMATE AND SOIL REQUIREMENT OF JUTE

Climate : In the northern hemisphere, cultivation stretches from 16°46′ N in Burma to 27°28′ N in India. The most efficient area in Bangladesh located between 58.5°E and 92°E and between 22.5°N and 26.5°N. The climate during jute season in Bangladesh, characterised by moderate rain fall and warm humid atmosphere, allow jute raising with low investments.

Rainfall : The average annual rainfall is 200 mm in Bangladesh, between 1500 and 2000 mm in jute areas in eastern India and between 1800 and 2800 mm in the main jute area in the Pegu -Irrawady region of Burma.

The total amount of rain and its distribution bears on the various intercultural operations and also on the growth and yield of jute. The entire cycle is divided into three distinct phases.

a) Nor-wester or pre-monsoon shower.

b) Lull or dry period between norwester and monsoon.

c) Monsoon or wet period.

The first phase helps crop initiation and stabilization and governs the main period of growth.

In the ideal situation pre-monsoon rain starts about the middle of March and the amount received is 100-120 mm spread over 15 to 20 days. Optimal spread and amount of rainfall depends on the soil and to some extent topography. The rain should be just enough to permit land preparation, sowing and germination.

Light rainfall occurs at intervals of 10 days or so, to keep the seedlings growing yet permit the two essential intercultural operations of thinning and weeding. If the period of lull stretches beyond a month, the jute crop tends to wilt. During this period of lull in rains , the thinning and weeding of the crop is completed and the seedlings become tall and strong enough to withstand the regours of monsoon. Incessant rain or waterlogged conditions are definitely harmful. Whereas, the *capsularis* varieties can stand waterlogging to some extent during the later stage of growth, *olitorius* varieties cannot.

The monsoon rain should be between 1200 and 1500 mm of precipitation distributed over 75 to 80 days. Strong sunshine and moderately strong wind alternating with rains are conducive to good growth. Continuous cloudy sky, prolonged drizzle or frequent heavy rain are not ideal. A crop sown during the monsoon or in a situation where weeding or thinning are required during that phase seldom gives either a uniform crop or high yield. Most of major jute growing South Bengal districts in West Bengal is considered the most efficient producer of olitorius jute in India.

In Bangladesh, considered the most efficient producer of jute under rainfed condition, the rainfall pattern along the Mymunsingh, Dhaka, Faridpur axis the norwester is regularly followed by a relatively dry spell, the monsoon is seldom in excess.

Shifts in the beginning and ending of the rainfall cycle mean early or late sowing. In the later case, yield would be adversely affected.

Shower in excess may not harm the crop, in sandy soils with a low water table. They do retard growth particularly when water stagnates in the fields.

Jute is a crop which thrives with rain and sunshine. The Indo-Bangladesh jute belt, which has a natural phenomenon of the right kind of rain and sunshine, continues to be the major producer of jute with lowest investment. Rain below 1000 mm is not helpful. Alternate sunshine and rainy days are most conducive to growth. The dry spell proceeding the break of monsoon enables the essential weeding and thinning operations to be carried out.

Relative Humidity (RH) : If a young crop, during the dry phase of April or May is subjected to low humidity at or below 19% for a prolonged period, premature flowering is induced in many of the varieties. This premature flowering occurred in all most all the jute growing districts irrespective of the varieties planted.

There are reasons to believe that the activity of the secondary cambium that produces the fibre has a rhythmic relationship with the variation in R.H. But the exact correlation can be established only through careful planned study.

Temperature : The jute plant requires a warm humid climate to grow into a tall fiber crop. It survives between 17°C and 41.5°C, thrives at 34°C ± 2°C and refuses to grow when temperature exceeds or falls below this range. Low temperature induces flowering.

Soil minimum and soil maximum temperature at 5 cm depth during March are between 19.7°C and 38°C, respectively. During April and May, the maximum toleration is 42.4°C and during monsoon, soil temperature is reduced to a usual range of 27.1°C to 34°C.

Day Length : Jute is a short day plant. In northern hemisphere, known cultivated varieties sown in summer start in late August or early September and come to 100 percent flower in 15 to 20 days. Genotypes which flower early or late also exist.

Recently a variety evolved from equatorial *C. Capsularis* has produced winter jute in commercial quantities. The variety was sown in November in a place located at 16°N in the Irrawaddy division and when harvested in early April gave 2.4 t ha^{-1} yield. The temperature in this tract was within the range of 19.7°C and 33.5°C.

Soil :

The Indo-Bangladesh Region :

The soil of the Indo-Bangladesh region on which jute is cultivated can be classified under 3 broad groups, new or grey alluvium, alluvium transported from lateritic areas and red and or lateritic soils. Soils with a low pH give a poor crop, the optimum pH being around 6.4.

i) New or Grey Alluvium : This soil is a gift of the river systems flowing from the Himalayas, including Ganges, Teesta, Bramhaputra and their tributaries. Fresh deposits of silt are laid wherever land is reached by flood or fluvial action; such soil is extremely fertile. In those regions of Bangladesh, North Bengal and Assam, which are annually flooded, 1 cm acre soil contains roughly 4 kg N (of 10% availability). Farmers in these areas use little fertilizer yet produce 1.5 to 2.2 t ha^{-1} of best quality fibre with native fertility.

At least 87% of the total jute crop of the Indo-Bangladesh belt is raised on grey alluvium. The soil in Nepal, North Bengal and the North bank of Bramhaputra ranges from sandy loam to loam and slightly acidic with pH ranging between 5.8 and 6.2. In such soil , besides correction of acidity, judicious use of manures and fertilizer is necessary to have a favourable yield.

In loam and clay loam soils with pH range from 6.2 to 7.2, good jute is produced wherever drainage is adequate. Clay soils are not very suitable for jute cultivation. The innumerable variation in soils due to topography, rainfall and inundation call for varied management practices.

ii) Alluvium Transported from Lateritic Areas : The second type of soil, the alluvium transported from lateritic areas, is represented by jute soils in the delta of Mahananda, Bramhani and Baitarani which flow through regions of lateritic and metamorphic rocks and therefore, have a certain affinity with secondary laterite. The soil, consisting of clay and silt in different proportions, is deficient in Ca, Mg and N. The pH in most areas is between 5.4 and 6.2. For a successful jute crop, liberal manure and fertilizer application is necessary. Liming enhances utilization of nutrients where soil has a low pH particularly when the source of lime is dolomite. In places, the clay component is more and stimulates in many respects lateritic clay; in such soil, only a poor crop can be sustained. In canal fed areas in the Cuttack district where the soil has attained a loamy texture, good jute is grown with average management skill.

iii) Red Soil : The third category is the red soil, as seen in the State Agril. Farm, Dhaka. Red soils of similar kinds occur in some part of Rajsahi, Burdwan, Madhupur tract of Bangladesh and also in the Cuttack – Balasore area in Orissa. In such a soil, it is difficult to carry out intercultural operations except when the soil has the right amount of moisture. In general, all red soils are deficient in N and without manure and fertilizer application, it is not possible to raise a good crop of jute. The other two major elements, P and K are also low. The pH is found to be within the range of 4.8 to 5.8, but continuous application of lime induces depletion of K. Jute in red soils require heavy manuring and balanced application of N, P and K as well as cautious application of lime.

Jute can be grown on varieties of soils, but sandy and clayey soils are difficult to work with low pH, leaching and water logging hamper growth and incite certain diseases. Jute requires a soil that yields to the plough, is easily pulverised and has a pH within the range of 6.2 to 7.6, the ideal being 6.4. The soil should be of good depth and the moisture retention capacity must be medium to high. Jute starts wilting when soil moisture falls below 12% and such wilting is irreversible when moisture drops to 6%. Soils with high organic matter content respond better when fertilizers are applied. The most suitable soil belongs to the river silt systems. Commercial success depends on the right combination of soil, rainfall, temperature and skilled management.

Climate and Soil for Mesta Cultivation :

Fibre of mesta is obtained from stem of *Hibiscus cannabinus* (H.C.) and *Hibiscus subdariffa* (H.S.). H.C. mesta is quick growing, possesses stiffer stem and flowers in late September to October, while H.S. mesta has flexible stem, slow growing and flowers in mid-November irrespective of sowing dates. Warm humid climate suits both H.C. and HS mesta but both grow in dry rainfed areas, the later having better drought resistance property. In areas with 500-900 mm rainfall, HC suits better by virtue of shorter duration and faster growth. Heavy, continuous rain and low temperature are harmful. Neither of them can stand prolonged waterlogging. Both are *kharif* crops and sown in April-June with the first showers of the monsoon.

Mesta is grown on a wide range of soils including new and old alluvium and lateritic loam. It thrives well even without much care. Soil pH 6.0-7.5 is suitable for mesta. Best crops is obtained in friable soils rich in organic matter. With a high pH (above 7.0) of soil, chlorosis is induced in HS mesta due to iron deficiency. The rich loam gives the highest yield, both grow on a variety of soils including new and old alluvium; but acid soils are not suitable without amendment. Both types of mesta are unfit for low-lying areas subject to inundation.

□□□□□
□□□

CHAPTER 3

AGRONOMIC MANAGEMENT OF JUTE CROP

In many low lying and fringe areas of horse show lakes and beels, jute has to be sown in end February to early part of March, so that it can be harvested early with the first on set of heavy shower by middle of June when such areas get permanently submerged. Capsularis is the ideal variety to such situation.

In terai zone, there remains a considerable amount of residual soil moisture in soils having a favourable profile. Pre-monsoon showers may again supplement such moisture. These two factors again offer scope for early sowing of jute in medium land situation also. Here again, Capsularis is the ideal variety to suit.

Under normal cropping sequence jute is followed by transplanted rice. Capsularis jute can be sown at least a fortnight ahead of Olitorius jute and can be harvested early. Capsularis jute is better-suited type in a jute-rice sequence.

Even the introduction of so called Chaitali tossa variety is during second half of March (month of Chaitra) may be disastrous for terai region. During May if there is continued shower overcast skies for a couple of days the right temperature may drop below 20°C and this may encourage premature flowering in olitorius.

In medium low land situation there may be early waterlogging after a moderate heavy shower. This waterlogging situation is not at all suitable for olitorius jute.

But defying all the above favourable factors for Capsularis cultivation, there has been a sudden spurt in favour of olitorius jute and the replacement has already been to the tune of more than 80 p.c. The favourable considerations are higher productivity and much better market price. Another point in favour of growing more and more olitorius jute in Terai is the Veg./potato –jute sequence. The replacement is justified, but it should have been limited to the tune of 50 p.c. covering the medium high and high level situations. Such division would also favour in distribution of peak weeding season, as Capsularis and Olitorius move at least 15 days apart in all aspects of cultural practices.

The cultural practices, which should be taken into consideration, may be summarized as follows.

Land Preparation :

Land preparation should have the following objectives in view:

i. Jute seed is very fine and fine tilth is necessary.

ii. Jute seed is highly sensitive regarding need of soil moisture – correct soil moisture regime should be provided.

Knowing the above fact, very well farmers pay special attention to prepare seed bed for jute. The land is ploughed and cross ploughed 6-8 times with country plough associated with necessary laddering. If power tillers are used three ploughings may be enough.

As moisture has to be assured, level preparation is initiated only after pre-monsoon shower.

If irrigation is available pre-sowing irrigation should be provided during final stage of land preparation if situation demands so.

Finally all stubbles are removed and soils are trimmed and the next step of sowing is adopted.

Sowing :

a) Seed Treatment :

Seeds should be treated with mercurial compounds like Ceresan at 300:1(seed : chemical). Dithane M-45 at 3.4 gms lit^{-1} of water may be used for wet treatment. The seed is immersed for 5-6 minutes and dried in shed before sowing.

b) Time of Sowing :

i) Early Sowing : In low areas where there is a chance of inundation by sudden rain or flash flood sowing may be done during end February to first week of March with residual soil moisture.

ii) Normal sowing : Normal sowing in case of Capsularis may start from early March and continue upto end of April. In earlier days normal sowing with Olitorius jute normally starts from middle of April. But with the development of varieties like JRO-878, JRO-7835, JRO-524, JRO-8432 sowing is possible from middle of March, but in areas where there is a possibility of drop in minimum temperature below 20^{0}C, there may be pre-maturing flowering. The problem is more common in North Bengal areas. Hence, sowing of Olitorius jute should be delayed upto first week of April. But in South Bengal condition sowing of these varieties can be taken up after mid March.

c) Method of Sowing :

There are two major means of sowing. Broadcasting is the common method. But considering the cost of weeding, line sowing should be advocated. This will permit wheel hoe operation in between rows several times for effective low cost weed control.

Spacing : Row to row spacing : 25 cm. for Olitorius
30 cm. for Capsularis
Plant to plant spacing : 5-7 cm.

Seed rate	Seed rate in kg ha^{-1}	
	Broadcasting	Line sowing
Capsularis	10 kg	7 kg
Olitoorius	7 kg	5 kg

Manuring :

Principle of manuring in jute : Jute is a crop which needs moderate level of manuring. But before fixing schedule of manuring few points have to be considered. The points are:

i. Initial fertility status
ii. Cropping intensity
iii. Nature of preceeding crop.

When initial fertility is poor, level of fertilization should be above average. In case of higher intensity of cropping and preceeding crop of exhaustive type, level of manuring should be high. Jute after potato may be highly benfited from the residual level of manuring.

Level of manuring	Kg ha^{-1}		
	N	P	K
Capsularis	40-80	20-40	20-60
Olitorius	20-60	20-40	20-60

Some points regarding manuring of Jute :

i. Yield of Olitorius is higher, but this species can do better with less amount of nitrogen compared to Capsularis.

ii. In many cases Capsularis continue to register higher yield upto 100 kg N ha^{-1}.

iii. Phosphorus and potassium have little direct effect on yield of fibre, but potassium has some effect on quality of jute fibre.

Liming Acid Soils :

Liming acid soil having pH below 5.5 often result in increase in crop yield. The same is true for jute. Liming of acid soils of North Bengal, Assam and Tripura in addition to providing direct benefit to the crop, reduces the chance of damage due to foot rot and stem rot diseases and finally enhances yield. Dolomite containing both calcium and magnesium and being available plenty in North Bengal and Bhutan border areas may be considered a superior soil amendment.

Applying Micronutrients :

Light soils in North Bengal, Assam and Tripura have been identified as deficient soil in micronutrient particularly in boron. Olitorius jute has been identified as more sensitive in boron deficiency. Application of 15 kg of boron per ha is recommended for better growth, yield and fibre quality.

Weed Control :

Weed control is considered to be very important intercultural operation in jute cultivation. Weed control bears two way significance in jute crop.

a) Weed control consumes lot of man days. In order to reduce cost weed control has to be completed at low cost.

b) Weeds must be controlled at early stage of the crop. Delayed weed control is not only more expensive, but permanently harmful too. The crop can hardly regain damage caused by delayed weed control.

i. Jute is grown during warm humid season.

ii. Jute is grown in medium high and high land situation.

iii. The soil texture is light in many cases.

The above factors favour high rate of weed growth and under North Bengal condition all the three favourable factors of weed growth combine together to form a formidable task of confronting weeds in jute timely.

Grasses, sedges and broad leaved all types of weeds are found in jute. Grasses are dominant groups followed by sedges and broad leaved weeds.

Weed Control Method :

Cultural - Pre-ploughing : Before actual sowing land may be fully prepared and left as such for one week. Some weeds will germinate and will be damaged when land is ploughed again and crop sown.

Weed evading practices : Germination of weed seeds + growth of weeds are usually more at a particular time. Seeds may be sown seven to ten days early. This allows good weather condition for effective and low cost weed control. Before weeds can over run the crop, the crop makes an effective canopy to smother the weeds if any as under growth.

Mechanical Means :

Mechanical means is considered to be most effective means of weed control. The following devices can be adopted.

i. Sowing the crop in rows well ahead of season.

ii. Operating wheel hoe in between rows three times at 4-5 days interval during bright weather conditions.

iii. Wheel hoe operation is followed by two hand weeding for removing weeds along the rows and also for thinning.

In case of broadcast crop, hand weeding remains the only means of weed control. At least two-hand weeding are necessary and for total weed control lot of man days may be consumed.

Chemical Methods :

Chemical method as such has not yet been very effective in terms of cost and effectiveness. Perfect selectivity and low cost is difficult to combine. But considering to never ending possibility of new organic compounds to come, some break through may be possible in near future. At present some chemical treatments along with one hand weeding have been found to be fairly effective.

1. **Fluchloralin (Basalin) :**

   1.5 kg a.i. ha^{-1} applied as pre-emergence spray 24-48 hrs. before sowing followed by one hand weeding at 21-25 days after sowing.

2. **Fusilade :**

   600 gms a.i. ha^{-1} applied as post emergence spray at 21 days after sowing applied as blanket spray.

3. **Thinning :**

   Thinning of jute is very important intercultural operation. During weeding initial thinning is done. But at 50-60 days crop age some plants may be found lacking in growth and fail to maintain height of the main population. Such plants remain thin undergrowth and should be removed to maintain quality of fibre. During years of good market price such plants are retted and fibres are extracted for family use and sale.

An optimum population of 5 (five) lakhs per ha is considered ideal for better yield and quality.

Management of Jute Pests :

Jute is liable to be infested by about thirty species of insects, two species of mites and nematodes. All the plant parts i.e. from root to capsules are attacked by these pests causing both quantitative and qualitative losses. However, amongst these pests, jute apion, semilooper, indico caterpillar, Bihar hairy caterpillar and Yellow mite are considered as major ones and appear almost regularly while, others are of minor or sporadic importance.

In perspective of present day situation, pest management of jute should be formulated considering socio-economic as well as agro-ecological background of the region concerned, and thus approach for integration of pest management strategies comes into the force.

Problems in Jute Pest Management :

i. Jute as well as bast fibre crops perhaps the taller, annual field crop producing above canopy which not only difficult for pesticidal applications but also create favourable microclimate for multiplication of pests and diseases.

ii. It is difficult to apply pesticides evenly on the entire canopy particularly at the apical portions which is the zone of two most important pests of jute namely, jute semilooper and yellow mite.

iii. Jute growing in pre-kharif season, often faces frequent rain which is more common in terai region. This causes washing of pesticides after spraying, that not only lowers the concentration of pesticides favouring resurgence of pest population but also creates pesticidal contamination or hazards to the water resources or water bodies through percolation to the water table as well as by the surface run-off.

iv. Because of dense crop canopy pest surveillance programme becomes difficult to test.

v. Lack of systematic work on conservation and utilization of natural enemies of jute pests.

vi. Problems of root-knot nematodes in unique soil climate conditions in terai region under changing cropping sequence particularly jute-potato rotation appears as an important bottleneck in the jute production. Expansion of olitorius jute without considering land situation creates another problem in jute growing. All these cause heavy outbreak of pests and diseases of jute in terai region of West Bengal.

In view of unique problems of jute pest management research findings should be integrated so far achieved in this region in the following ways in order to get a technically feasible, ecologically sound and economically viable pest management programme.

I. Pest surveillance :

i. Regular survey of jute field to ascertain pest population before adopting any control measure particularly through chemical means.

ii. Collection and destruction of affected seedlings by jute apion at the time of thinning in order to destroy the source of infestation at the early stage of crop growth.

iii. Collection and destruction of egg masses and the early stage gregarious larvae on leaves of Bihar hairy caterpillar.

II. Use of climatological and survey data for forecasting of pest outbreak and thereby to take decision for control measures :

i. **Jute semilooper (*Anomis sabulifera*) passes three broads (generations) from end of June to mid August of which 2nd broad is more damaging. Hence, preventive measures during end of June may subscribe the population at succeeding generations (broads). The green semilooper (*Ilattia octo*) a polyphegous pest has been recorded to damage the tender foliages of olitorius jute in some areas.**

ii. Jute apion favours Capsularis jute more than Olitorius. Early sown crops are liable to damage more than late sown crop. Applications of high doses of N-fertilizers increases the infestation.

iii. Occasional light rains which keep the relative humidity high for a prolong period is most favourable for yellow mite of jute. Olitorius jute is preferred more than Capsularis jute. In terai region yellow mite appears in peak population during June-July. So, decision for adopting control measure will be taken accordingly.

iv. **Red mite favour Capsularis jute whereas, yellow mite is more serious on olitorius jute. Prolong dry spell during growing season favours its multiplication. Red mite damage the older leaves but the yellow mite damage the apical leaves.**

v. **Infestation of root knot nematodes (*Meloidogyme spp.*) can be escaped by delayed sowing of jute after middle of April. Galls are formed on the roots; plants become stunted; apical growing points may dry up.**

III. Conservation and use of natural enemies :

Bihar hairy caterpillar (*Silosoma obliqua*) and semilooper are two important and major pests of jute in terai region, though appear in large number in early instars but population is not so proportionately high in late instars when they become voracious feeders. This is due to parasitisation by two braconial larval parasites, *Apantales ruficrus* (on Bihar hairy caterpillar) and *Doryctes spp.* (on jute semilooper). They parasitise upto a tune of 2-90% and 8-100% of the larval respectively. Hence, waiting to take pesticidal measures till late stage will not only save the huge cost of pesticidal application but also will favour conservation of natural enemies. Indigo caterpillar (*Spodoptera exigua*) and the cricket (*Brachytrypes protentosus*) are essentially early season pests of the crop. Indigo caterpillar is widely distributed causing heavy damage to early sown crop under hot and dry condition. The pest is active from March-June but activity declines during rainy season. Early instar larvae web up the apical leaves while grown up larvae remain hidden under various shelters during light hours and feed voraciously during night.

Adults of *Myllocerus discolor* damage the young leaves of olitorius jute during May-June. Egg, larval and pupal stages could be found in soil. Infestation declines after the on set of monsoon.

Jute stem girdler (*Nupserha bicolor postbrunnea*) is a minor pest of olitorius jute. Well-managed healthy plants are not generally infested.

The mealy bug (*Ferrisia virgata*) may appear as a minor pest on the seed crops of olitorius jute. Damaged plants produce fewer number of deformed pods.

IV. Chemical measures :

Unnecessary use of pesticides should be avoided. Pesticides are harmful to natural enemies as well as costly too. Insecticides should be applied in proper dosages and manner. Following pesticides may be used for different kinds of pests.

Indigo caterpillar, Semilooper and Hairy caterpillar: spraying with monocrotophos, endosulfan or fenitrothion at 0.05% a.i. or cypermethrin at.01% a.i.

Jute Apion : spraying with 0.5% a.i. of endosulfan or fenitrothion or 0.1% carboryl.

Mealy Bug : Spraying with 0.05% dimethoate or endosulfan.

Yellow Mite : Spraying with wettable sulphur dicofol or ethion.

Nematode : In heavily infested soils carboferron 3G at 1 kg a.i. ha^{-1} in soil.

V. Management of root-rot nematodes :

Application of poultry manure to jute paddy rotation is the best way of root rot nematode management, so far crop growth, fibre yield, suppression of nematode infestation and cost are concerned. The application of carbofuran 2kg a.i. ha^{-1}, twice on jute in the above rotation is the best way for their rapid suppression.

Management Diseases of Jute :

Important disease of jute are stem rot, anthracnose, bacterial blight, bacterial wilt and root knot. The causal organisms as well as their control are summarised below.

A. Stem rot :

Stem rot of jute caused by *Macrophomina phaseolina* is a serious disease of the crop in West Bengal, Assam, Orissa, Bihar. Both the quality and its out-turn are affected by this disease.

Symptoms : Plants in all growth stages are attacked by this pathogen. On seedlings, the earliest symptoms appear on the hypocotyls and cotyledons in the form of blackish-steaks . In younger seedlings damping off is common, if the weather is moist.

On the developed seedlings or young plants dark brown lesions develop on stem particularly towards the base of the plant resulting dropping of leaves and death of seedlings. Death of seedlings creates gaps on the field. Jute as a rule, is thickly sown as to minimise branching of plants.

Gaps induce branching of plants :

On the stem of grown up plants, dark brown lesions develop at any point which increase in size, several such lesions coalesce and girdle the stem. Girdling stem causes death of upper part of the plant where the steaks run along the length of the stem without girdling, the cortea becomes shredded, exposing fibre tissue.

Occasionally, roots are infected and the plants show wilting symptoms. Infected roots show rustly brown discolouration as compared to light yellow healthy roots.

On the leaves of mature plants necrotic lesions appear at the apex, along their margins, but later the entire leaf-blade, mid ribs and petioles are completely involved. On capsule dark brown lesions of different size are formed. On close examination of the lesions on stem, leaf and fruits small dot like, numerous raised pycnidia are found.

Source : In majority cases seeds carry the pathogen in sclerotial or pycnidial form and disease is intiated by seed borne inoculation. The pathogen can also survive in soil. Conidia produced in pycnidia on the developing seedling help in spread of the disease to the stem and leaf. Lean and thin plants within the normally grown plant population are most susceptible. These plants help in build up of inoculum.

Control :

i. Spraying of Carbendazim (Bavistin , J.K. Stein, Derosal etc. 0.1%) in the seed producing field prevent seed infection.

ii. Seed treatment with fungicides (Thiram, Captan 4g kg^{-1}, Carbendazim 2g kg^{-1}) prevents seedling infections.

iii. Application of lime is reported to reduce disease incidence.

iv. Wherever necessary spraying of copper fungicide (0.4%) or Carbendazim (0.1%) can be followed for protection of stem and leaf infection.

v. Destruction or removal of crop residues after harvest of the crop.

vi. Growing of transplanted rice after harvest of jute crop residues, the scope of soil survival of the pathogen.

Impact of the Diseases on the other Crops :

The fungal pathogen *Macrophomina phasevlina* can infect most of the vegetables and pulse crop. Growing of vegetables and pulse crop after jute in upland situation enhances the scope of survival of the pathogen and vegetable crops may be severely affected by this pathogen. Proper care should be taken in selection of crop after jute.

B. Anthracnose :

The disease is caused by two species of *Collectotrichum. Collectotrichum corochorum* infects *C. Capsularis and Collectotrichum gleosporiodes* infects *C. olitorius*. The former is prevalent in Assam and North Bengal and the latter is common on occurrence in all olitorius growing areas.

Symptom : Both the pathogens mainly infect stem of mature plant producing dark brown lesion. In the Capsularis jute the dark lesions rapidly coalesce forming a big slightly dark path. It may cause drying of upper part of the crop and the plant may break. In olitorius jute, the lesions are small spreading over the stem and are slightly depressed in nature. Occasionally the spots coalesce. Fibre quality of the affected plants becomes poor.

Source : The disease is seed borne.

Control : Seed treatment with fungicides and spraying of copper fungicides, if necessary are suggested for control of the disease.

C. Bacterial wilts :

The disease is caused by a bacteria, *Pseudomonas solanacearum* infecting mostly olitorius jute.

Symptoms : Usually mature plants show the disease symptoms. Initially leaves of the affected plants droop showing temporary wilting symptom. Within 3 to 5 days the plants finally wilt and die. Infected plants simultaneously may be infected by stem rot and root knot pathogens.

To confirm the bacterial nature of the disease take a small piece of stem from the basal part and put it in a glass of clean water. After a few minutes white bacterial core will come out from the cut end of the stem.

Source : The disease is soil born. After the death of the affected plants the bacterium survives in soil. Same bacterium infects all the solanaceous vegetables and infection may come from such crops. Incidence of the disease is more when jute is grown after potato.

Control : Crop rotation with maize, sorghum or other cereals, mustard, reduces bacterial population in soil. Chemicals application is ineffective.

D. Root knot :

The disease is common for both type of jute, particularly in sandy loam, loam soil in upland situation.

Generally jute after infection of root knot nematode do not show any external symptoms. Growth appears to be normal. Severely infected plants show unthiftly development and stunted growth. The leaves are yellowish green to yellow tend to droop. Sudden wilting may be seen, if the wilt organisms are associated.

Most stricking symptoms are found on root knot after uprooting of the plants. The main and lateral roots bear spherical to elongated irregular galls.

Source : The nematode pathogen (*Meloidogyne* sp.) survive in soil.

Control : Crop rotation with cereal is recommended. Application of oil cake and saw dust reduces nematode population.

Impact of Disease on other Crop :

This nematode has wide host range. Most of the vegetables are infected by this pathogen. Growing of vegetables immediately after harvest of nematode infected jute crop should be avoided. Incidence of this nematode increases severity of the bacterial wilt disease in jute and other crops. Complex association of *Macrophomina phaseolina*, *Pseudomonas solancearum* and root knot nematodes become responsible for death of large number of jute plants. Proper sanitation, crop rotation should be followed in growing in such field.

Harvesting of Jute :

Jute may be harvested at any time between 90 to 150 days of crop age. Early harvest gives better quality fibre but low yield, while late harvest gives higher yield but coarse fibre. To compromise between quality and quantity, the harvesting is recommended at early pod stage or at the age of 110 to 125 days. The other consideration for early harvest is to accommodate rice transplanting in accordance with cropping pattern chosen.

In general, harvesting is done at full maturity when seed pods become brown. In *Corchorus capsularis* good quality seed is obtained by harvesting when stems and pods are not dead ripe. But for *Corchorus olitorius* it should be done when the stems and pods are dead ripe.

Harvesting is done by cutting plants at ground level by a sharp sickle during the morning hours and the plants are kept on the field for 2 to 3 days for shedding the leaves. Next, the plants are tied into bundles of 20 to 25 cm. diameter and the tops are chopped off in the field. Bundles are also made immediately after harvest and the bundles are kept standing in the field to shed the leaves. During the period, when the harvested plants are left in the field the tissues shrink and cells rupture. This facilitates the entry of microorganisms into the stems while retting.

Which Type to Grow :

Two species of jute are available for cultivation e.g. *Corchorus capsularis* and *Corchorus olitorius*. Of the two C. capsularis is hardy in nature and can be grown both in high and low land and have a more diversity in sowing time. On the otherhand, C. olitorius is suitable for well drained high land only. But from economical view points, olitorius offer more attractive qualities in the form of better yield, better fibre quality and more market price. Thus from the same investment the return from olitorius is more than capsularis. Farmers prefer growing olitorius for obvious reasons. Capsularis offer better scope for early sowing, timely weed control during better sunny days, evading possibility of being water-logged and damaged as may happen to late sowing of olitorius in medium low land. Because of the mild thermo-sensitivity properly sown even in April may show premature flowering with drop in temperature below 25 ^{0}C, after a few days of rains. Hence for North Bengal both the species can be accommodated – capsularis for early sowing in low lying areas and olitorius for delayed sowing in high land areas. This combination would also ease the pressure of cultural practices all at a time.

Varieties :

Some good varieties of both the capsularis and olitorius are available which may be offered to farmers. But current achievement in releasing new varieties capable of recording 20-25 p.c. higher yield in comparison with the existing or varieties are not forthcoming. A genetic break through is now essential to combat the problem of non-remunerative nature of jute crop.

***Corchorus Capsularis* :**

Sabujsona (JRC – 212) : released during 1954.

Parentage : Straight selection from an indigenous type.

Stem and Foliage colour : Full green.

Seed coat colour : Shining chocolate.

Fruit : Non-shattering.

Nature of flowering : Mid March sowing does not include premature flowering.

Days to 50% flowering : 150-160 days.

Seed to seed maturity : 220-230 days.

Optimum sowing time : Mid March to Mid April.

Suitable land for sowing : Mid and high land.

Average fibre yield : 25-28 quintal ha^{-1}.

This variety is suitable for northern as well as southern districts of W.B. and for other states as well.

Sonali (JRC 321) : released during 1954.

Parentage : Straight selection from an indigenous type 'Hewti'.

Stem and foliage colour : Stem and upper surface of leaf petiole coppery red.

Seed coat colour : Shining chocolate.

Fruit : Non shattering.

Nature of flowering : Late February sowing does not induce premature flowering.

Days to 50% flowering : 130-140 days.

Seed to seed maturity : 200-210 days.

Optimum sowing time : Late February to late March.

Suitable land for sowing : Low and mid land.

Average fibre yield : 20-25 quintal $hectare^{-1}$.

This variety is ideal for North Bengal region. It is equally suitable for Assam, eastern UP.

Shyamali (JRC 7447) : released during 1971.

Parentage : X-ray derivative of JRC 212.

Stem and foliage colour : Full green.

Seed coat colour : Shining chocolate.

Fruit : Non-shattering.

Nature of flowering : Mid March sowing does not induce premature flowering.

Days to 50% flowering : 150-160 days.

Seed to seed maturity : 225-235 days.

Optimum sowing time : Mid March to Mid April.

Suitable land for sowing : Low and mid land.

Average fibre yield : 28-30 quintals ha^{-1}.

Baladev (JRC 4444) : released during early eighties.

Parentage : Straight selection from an indigenous type.

Stem foliage colour : Full green.

Seed coat colour : Shining chocolate.

Fruit : Non shattering.

Nature of flowering : Mid. March sowing does not induce premature flowering.

Days to 50% flowering : 150-160 days.

Seed to seed maturity : 220-230 days.

Optimum sowing time : Mid March to mid April.

Suitable land for sowing : Mid and high land.

Average fibre yield : 25-28 quintals ha^{-1}.

Reshma (UPC 94) : released during 1983.

Parentage : A selection from a cross between JRC 321 and JRC 212.

Stem foliage colour : Stem and upper surface of leaf petiole coppery red.

Seed coat colour : Shining chocolate.

Fruit : Non shattering.

Nature of flowering : Late February sowing does not induce premature flowering.

Days to 50% flowering : 140-150 days.

Seed to seed maturity : 210-225 days.

Optimum sowing time : Late February to late March.

Suitable land for sowing : Low and mid land.

Average fibre yield : 25-27 quintals ha^{-1}.

Padma (Hybrid C) : released during 1983.

Parentage : A selection from a cross between JRC 6165 and JRC 412.

Stem foliage colour : Stem and upper surface of leaf petiole coppery red.

Seed coat colour : Shining chocolate.

Fruit : Non shattering.

Nature of flowering : 140-150 days.

Seed to seed maturity : 200-215 days.

Optimum sowing time : Late February to late March.

Suitable land for sowing : Low and mid land.

Average fibre yield : 25-28 quintals ha^{-1}.

Shrabanti white (JRC 698) : released during 1999.

Parentage : Selection from a cross involving 5 indigenous and 8 exotic parents.

Stem foliage colour : Plant full green.

Seed coat colour : Blackish grey.

Optimum sowing time : Mid. March.

Suitable land for sowing : Low and mid land.

Average fibre yield : 30 quintals ha^{-1}.

***Corchorus Olitorius* :**

Baishakhi Tossa (JRO 632) : released during 1954.

Parentage : Straight selection from an indigenous type.

Stem and foliage colour : Full green.

Seed coat colour : Brownish grey.

Fruit : Shatter on maturity and drying.

Nature of flowering : Earlier than mid April sowing induces premature flowering.

Days to 50% flowering : 130-140 days.

Seed to seed maturity : 200-240 days.

Optimum sowing time : Mid. To end of April.

Suitable land for sowing : Mid.and high land.

Average fibre yield : 30 quintals ha^{-1}.

Chaitali Tossa (JRO 878) : released during 1967.

Parentage : Selection from a cross between JRO 632 and Sudan grass.

Stem and foliage colour : Full green.

Seed coat colour : Blackish grey.

Fruit : Does not shatter on maturity and drying.

Nature of flowering : Mid March sowing does not induce premature flowering.

Days to 50% flowering : 180-200 days.

Seed to seed maturity : 260-280 days.

Optimum sowing time : Mid. March to mid April.

Suitable land for sowing : Mid and high land.

Average fibre yield : 32 quintals ha^{-1}.

Basudev (JRO 7835) : released during 1971.

Parentage : Selection from a cross between JRO 632 and Sudan grass.

Stem and foliage colour : Full green.

Seed coat colour : Blackish grey.

Fruit : Does not shatter on maturity and drying.

Nature of flowering : Mid March sowing does not induce premature flowering.

Days to 50% flowering : 180-200 days.

Seed to seed maturity : 260-280 days.

Optimum sowing time : Mid. March to mid April.

Suitable land for sowing : Mid and high land.

Average fibre yield : 34 quintals ha^{-1}.

Navin (JRO 524) : released during 1977.

Parentage :Selection from a cross between JRO 632 and Sudan grass.

Stem and foliage colour : Full green.

Seed coat colour : Blackish grey.

Fruit : Does not shatter on maturity and drying.

Nature of flowering : Mid March sowing does not induce premature flowering.

Days to 50% flowering : 180-200 days.

Seed to seed maturity : 260-280 days.

Optimum sowing time : Mid. March to mid April.

Suitable land for sowing : Mid and high land.

Average fibre yield : 36 quintals ha^{-1}.

Mahadev (TJ 40) : released during 1983.

Parentage : Fast Neutron derivative of JRO 632.

Stem and foliage colour : Full green.

Seed coat colour : Brownish grey.

Fruit : Shatter on maturity and drying.

Nature of flowering : Earlier than mid April sowing induces premature flowering.

Days to 50% flowering : 130-140 days.

Seed to seed maturity : 200-240 days.

Optimum sowing time : Mid. to end of April.

Suitable land for sowing : Mid and high land.

Average fibre yield : 35 quintals ha^{-1}(in jute belt of Orissa).

Savitry (JRO 3690) : released during 1985.

Parentage : Selection from a cross between tobacco leaf and long internode.

Stem and foliage colour : Full green.

Seed coat colour : Steel grey.

Fruit : Shatter on maturity and drying.

Nature of flowering : Earlier than mid April sowing induces premature flowering.

Days to 50% flowering : 130-140 days.

Seed to seed maturity : 260-280 days.

Optimum sowing time : Mid. to end of April.

Suitable land for sowing : Mid and high land.

Average fibre yield : 33 quintals ha^{-1}

Golden Jubilee Tossa (JRO 66) : released during 1998.

Parentage : Selection from a cross involving 3 indigenous and 3 exotic perents.

Stem and foliage colour : Full green.

Seed coat colour : Steel grey.

Fruit : Pod non-shattering.

Days to 50% flowering : 130-140 days.

Seed to seed maturity : 260-280 days.

Optimum sowing time : can be sown earlier April.

Suitable land for sowing : Mid. to high land.

Average fibre yield : 36 quintals ha^{-1}

Shakti Tossa (JRO 8432) : released during 1999.

Parentage : Selection of a cross of IC 15901 and Tanganyka - 1.

Stem and foliage colour : Full green.

Seed coat colour : Blakish grey.

Fruit : Does not ahatter on maturity.

Days to 50% maturity : 180-200 days.

Seed to seed maturity : 260-280 days.

Optimum sowing time : Mid. March to Mid April.

Suitable land for sowing : Mid to high land.

Average fibre yield : 32 quintals ha^{-1}.

Sowing :

Jute (*Olitoruis/Casularis*) and mesta are rainfed crops for small and marginal farmers , grown with the help of pre-monsoon showers received in April and May. Sowing is often delayed due to non/late receipt of rainfall. *Olitorius* jute is cultivated in about 90% area whereas *Capsularis* is limited to low land areas of North Bengal and Assam. Andhra Pradesh accounts for more than 50% of area and production of mesta. Premature flowering resistant Olitorius varieties like JRO-878, JRO-7835 and JRO-524, JRO-8432 can be sown from end of March whereas, *Capsularis* varieties are sown from end February. Mesta is normally sown in the month of May and June. Jute is normally harvested for fibre at 110 to 120 days crop age. Mesta is harvested at 150 to 180 days crop age. West Bengal is the leading producer of jute with about 73% area and 80% of production.

Seed Rate : Sowing is mainly done by broadcasting with only 10% area under line sowing. The seed rate of Capsularis jute is 7 to 10 kg; with *Olitorius* 5 to 7 kg and of Mesta 11 to 17 kg ha^{-1}.

Spacing : Plant to plant spacing in broadcast crop is maintained at 10 cm, whereas in line sown crop it is 5 to 7 cm. Multi-row seed drill evolved by **CRIJAF** could be useful for bringing in more area under row crop. Line sowing will increase fibre yield by 15 to 20%. Seed rate is reduced by 2 to 3 kg ha^{-1} as compared to broadcast. Line sowing also facilitates uniform germination, good crop stand, proper aeration and ultimately better fibre yield. Recommended row to row spacing is 25 cm. in *Olitorius* and 30 cm. in *Capsularis* and mesta.

Interculture Operations :

In broadcast interculture is done with *khurpi* which helps in loosing the soil and providing aeration. In a line sown crop wheel-hoe provides the required aeration and weeding in between the rows. Fertilizer application is also more uniform in a row sown crop as compared to broadcast.

❑❑❑❑❑
❑❑❑

CHAPTER 4 SOIL MANAGEMENT AND CROPPING SYSTEMS OF JUTE

Jute was affected due to continuous wetness. The desired effect due to fertilizer is often restricted under poor management. So continuous wetness and water logging are to be carefully avoided in jute fields for better utilization of the nutrients and profits from applied fertilizers.

Response to Application of Nitrogen :

Nitrogen gives universal response but the magnitude of response varies depending on several factors. Under rainfed conditions, the yield of jute increases by about 8-16 times per unit of nitrogen; with controlled irrigation and drainage facility, the increase may be even 15 to 25 times. Under control irrigation in areas of low rainfall the response to application of N is better than in areas with higher rainfall. The use of fertilizer on extensive scale in a limited sense gives better results than on intensive scale, that is instead of application of 60 kg N in one hectare, application of the same quantity in two hectares gives better yield.

Comparative Merits of different Nitrogenous Fertilizer on Jute :

From experimental results it has been found that application of nitrogen in general, increased the fibre yield significantly over control while ammonium sulphate proved the best source. The nitrates generally gave lower yields. Next effective source is urea.

Methods of Application of Nitrogenous Fertilizers on Jute :

It has been found that in general, application of nitrogen as top dressing yields better results than when applied as basal dressing, obviously due to heavier leaching losses in latter case; further a considerable portion of applied nitrogen in basal dressing is taken up by the weeds and the extra jute plants which are subsequently removed during thinning operation to maintain the required plant population.

Time and Method of N application under Rainfed Condition :

When sowing is undertaken with available soil moisture and drought prevails for sometime , application of the entire quality of nitrogen in a single dose just before sowing has been found to be economical.

Under Irrigated Condition :

Where drought was not a problem after sowing, full dose top dressed three weeks after sowing gave the highest fibre yield followed by basal plus half top dressing. In jute, deep placement of nitrogenous fertilizer did not prove beneficial. Thus, under irrigated condition for better results, nitrogenous fertilizers should be used as top dressing after weeding and thinning operation.

Foliar Feeding of Urea :

Foliar feeding of urea may be an alternative to soil application under condition when i) fertilizer is in short supply, ii) when soil application is difficult due to drought or excessive weed grown, iii) in areas where leaching is a severe problem.

10 to 15 kg N ha^{-1} as urea may be applied in two spray with 10-20 per cent urea concentration using low volume power sprayers. First spray should be done at 5-6 weeks crop age and the second 10-15 days after the first. Urea makes a compatible mixture with insecticides like thiodan, etc. As such, two sprays can be combined into one.

Response of Phosphorus on Yield of Jute :

Ordinarily, application of phosphorus has been found to have little effect on the yield of fibre in alluvial tracts of jute growing areas but in P deficient areas, it has given good response. Phosphate prevent lodging and improved the quality of fibre by balancing the quality ratio of fibre. Placement of 'P' at a distance of 5 cm away and 5 to 10 cm below the seeds, regulate the highest utilization of 'P' and thereby the uptake of 'P' by jute crop.

In acidic area sparingly soluble sources of 'P' viz. deflourinated phosphate, Ultrafas and Thomas slag have been found to be more effective than easily soluble sources.

Critical Value :

The critical value of 'P' in soil extracted by Olsen's methods when correlated with per cent yield was rated to be 23.4 kg P_2O_5 ha^{-1}.

'P' Fractionation :

The consistent relationship of aluminium phosphate fraction with all the dependent variables suggest that Al-P is a major source of P in jute growing calcium phosphate dominating soils.

Response of Potash on Yield of Jute :

Potash requirement of jute is quite high. In K-deficient tracts it increases yield when supplemented with N and P ; it also reduced the incidence of root rot and stem rot caused by *Macrophomina phaseoli*. On an average per unit of K, the yield increased by 2 to 5 times or even more. The soils of Assam may be categorised as responsive and those of 24-parganas district in West Bengal as non-responsive to K-fertilizer with respect to fibre yield. The 24 parganas were responsible for the insignificant response to K fertilization. Generally ,in very coarse textured soils more K is required.

Effect of Magnesium on The Yield of Jute :

Magnesium also plays an important role as a fertilizer for jute where this element is in short supply. In the acidic areas dolimitic lime stone or powdered basic slag may serve as better liming material than pure limestone. Dolomite contains fair quantities of magnesium and slag contains both phosphate and magnesium.

Relation between Potassium of Leaf and Disease Incidence :

A significant negative correlation (r = –0.646) between potassium concentration of leaf and percentage disease incidence was observed.

It was also observed that K/Ca ratio in leaf, bark and wood of *capsularis* jute was higher than the corresponding values of *olitorius* jute which makes the crop somewhat resistant to disease. It may also be stated that under comparable conditions, the *olitorius* is more susceptible to disease than the *capsularis* jute.

The incidence of disease is less in well drained soils. Experimental factors indicate that the i) uptake of K is more in early part of crop growth and ii) K/Ca ratio which remains higher becomes narrower in the later stages. This may be due to the nature of the crop. The another reason may be controlled availability of K due to alternate wetting and drying conditions prevailed in the jute fields. Well-drained and aerated soil condition helps in prevention of fixation of K and this probably explain the less incidence of disease in jute under such situation.

Effect of Micro-nutrient Application on The Yield of Jute :

In acid soils or where jute is continuously cultivated, the available contents of Fe, Mn, Cu and Zn, in soils at different depths show that with the increase in depth the available iron recorded as marked decrease in case of alluvial soil.

Jute being a fairly deep rooted crop derives nutrients not only from the surface layer but also from lower layers of the soils: Although the surface soils have been found to be adequately supported with available Fe, Mn, Cu and Zn, the soils at the lower layers have in many cases recorded values lower than the critical values suggested for other crops. This point should be taken into accounts while considering the micro-nutrient supplying power of soils to jute crop.

Application of B, Zn, Mo, Mn, Cu etc. with NPK in deficient tracts may give a boosting effect on the yield of jute. The deficient tracts need to be delineated, application of Fe, Zn and B has given positive trend of response at Nagaon in Assam, B, Mn and Zn at Sorbhog (Assam) and Cu, Mn and Zn at Bahraich, UP.

Iron manganese, copper and Zinc requirements for *Capsularis* jute is more than that of oiltorius jute.

General Recommendations for Jute :

1. Organic matter : 4 -7 tonnes ha^{-1} as FYM
2. Nitrogen : 20-60 kg N ha^{-1} for Olitorius
   40-80 kg N ha^{-1} for Capsularis
3. Phosphorus : 20-40 kg P_2O_5 ha^{-1}
4. Potassium : 20-60 K_2O ha^{-1}

5. Lime : 2-4 tonnes lime ha^{-1} wherever necessary
6. Magnesium : 20-40 kg MgO ha^{-1}

Mesta : Mesta is mainly grown in the states of Andhra Pradesh, Orissa, Bihar, West Bengal and also in Meghalaya, Assam, Tripura. Besides, mesta is also grown in Maharastra and Karnataka but mainly as mixed crop. The main mesta grown districts are Vijaynagar and Srikakulam in Andhra Pradesh. Besides, in other states a small area is covered under mesta.

In all these states nearly 0.19 million hectares of land are under mesta cultivation on which nearly 10.98 thousand bale of raw mesta (1 bale = 180 kg) are produced.

Mesta is generally sown on marginal soils with the on set of monsoon in May and June. It thrives even without much care. Soil pH 6.0-7.5 suitable for its production. The land should be prepared well before sowing. Nearly 10 tonnes ha^{-1} of FYM or compost may be mixed with soil at the time of preparation of land. Sowing of mesta should be done when the soil moisture is around 21 percent.

General Recommendation for Mesta :

1. Organic matter : 4-7 tonnes ha^{-1} as FYM
2. Nitrogen : 40-60 kg N ha^{-1} in two split doses; ½ dose at the time of sowing and the ½ dose between 4-6 weeks after sowing
3. Phosphorus : 20-40 kg P_2O_5 ha^{-1}
4. Potassium : 20-60 kg K_2O ha^{-1}

Nutrient Requirement for Production of One Quintal of Fibre :

	kg					Gm			
	N	P	K	Ca	Mg	Fe	Mn	Cu	Zn
C. Capsularis (JRC-212)	2.9	0.60	5.7	3.08	1.12	78.4	25.1	2.6	21.4
C. Olitorius (JRO632)	2.2	0.6	4.5	1.92	0.86	36.7	11.2	1.8	13.9
H.S. 4288	3.7	1.5	6.5	6.0	1.5	-	-	-	-
H.C. 867	3.2	1.3	6.3	5.8	1.4	-	-	-	-

Jute based Cropping Systems :

Normally jute is grown as mono crop when rain is well distributed. However, the availability of early, premature flowering resistant jute varieties and emergence of short duration water and fertilizer responsive high yielding varieties of different crops has made it possible to grow three crops in a sequence within one year where mono or even double cropping were, hither to, common practice.

Multiple cropping refers to a cropping system in which two or more crops are grown on a given area of land over a period of one year. It, thus ensures higher production per unit area per unit time and provides opportunities for more jobes and income through out the year. The risk of growing one crop is also minimised and often compensated by the other crop grown in the sequence.

A number of intensive crop rotations with jute were evolved at the Jute Research Institute and have been recommended for the adoption by the farmers. In all, 15 crop rotations (9 three-crop rotations and 6 two crop rotations) were conducted under three situations of water availability: a) adequate and assured irrigation conditions, b) limited and measured irrigation conditions c) normal rainfed conditions.

a) Cropping system under adequate and assured irrigation conditions :

Assured irrigation refers to the situation in which farmers have their own irrigation sources like shallow tubewells, pumpsets to provide irrigation water wherever needed. Under this conditions, demonstrations on following 5-three crop rotations were conducted: Jute-paddy-mustard, Jute-paddy-wheat, Jute-paddy-potato, Jute-paddy-cabbage, Jute-paddy-groundnut. The details of the agronomic practices adopted in these demonstration and the results obtained are presented in Table 1.

Of the five rotations conducted under adequate and assured irrigations, jute-paddy-cabbage produced the highest aggregate yield of 319 q ha^{-1} followed by 222.50 q ha^{-1} obtained from jute-paddy-potato rotations. While the lowest yield of 76.58 q ha^{-1} ha was recorded from a jute-paddy-mustard rotation. The higher yield of the two rotations were due to vegetable crops of cabbage and potato. The yield from jute-paddy-wheat rotation was 92.78 q ha^{-1} and that of jute-paddy-groundnut was 80.16 q ha^{-1}.

The net income of Rs. 18,284 = 00 ha^{-1} yr^{-1} was highest in Jute-paddy-cabbage followed by jute-paddy-potato (Rs. 17,983 ha^{-1} yr^{-1}) suggesting that introduction of vegetable crops in the rotations increased the income at farmers fields. The other two rotations such as jute-paddy-mustard and jute-paddy-groundnut fetched the returns of Rs. 13,356=00 and Rs. 13,328=00 ha^{-1} yr^{-1} respectively. The lowest net income of Rs. 12,786=00 ha^{-1} yr^{-1} was received by adopting jute-paddy-wheat rotations.

It is interesting to note that the cost benefit ratio of jute-paddy-groundnut was highest (1.83) followed by jute-paddy-cabbage (1.67) and jute-paddy-mustard (1.65) mainly due to their higher selling prices.

b) Cropping system under limited and unassured irrigation conditions :

Under this situation, farmers do not have their own source of irrigation but depend on the highered irrigation facilities. When water supply is limited and unassured, 2 to 3 irrigations are provided at critical stages meeting only 2/3rd to 3/4th of the water needs of crops. Following 4 three crop rotations were conducted under this condition :Jute-paddy-mustard, Jute-paddy-pea, Jute-paddy-tomato and Jute-paddy-lentil.

Among four three crop rotations conducted under limited and unassured irrigated conditions, jute-paddy-tomato produced the highest total yield of 147.50 q ha^{-1} while jute-paddy-lentil gave the lowest yield of 73.40 q ha^{-1} .The second highest yield of 115.83 q ha^{-1} was received from jute-paddy-pea. Jute-paddy-mustard secured third position in respect of total yield (75.31 q ha^{-1}).

The maximum monetary net income of Rs. 16,567 ha^{-1} yr^{-1} was obtained by adopting jute-paddy-pea rotation followed by jute-paddy-tomato which gave Rs. 15,713 ha^{-1} yr^{-1}. Both pea and tomato were sold for green vegetables purposes. The rotation of jute-paddy-mustard gave a net income of Rs. 13,012 $ha^{-1}yr^{-1}$ while maximum net income of Rs. 12,964 ha $^{-1}yr^{-1}$was realised from jute-paddy-lentil.

The cost benefit ratio of jute-paddy-pea was highest (1.82) because of higher return from the sale of green pea followed by jute-paddy-tomato (1.70). The cost-benefit ratio was almost similar for jute-paddy-lentil (1.63) and jute-paddy-mustard (1.64).

c) Cropping systems under rainfed conditions :

Under normal rainfed condition, 4 two crop rotations conducted were: Jute-paddy, Jute-pea, Jute-soyabean and Jute-lentil. The purpose of such demonstration was to encourage farmmers to grow second crop after jute under upland and midland by utilizing subsoil or residual moisture at the sowing time and availing winter rain during growing period.

The details of the package of agronomic practices followed and varieties used for different crop rotations and the results obtained are shown in Tables 1-3.

Of the five rotations conducted under normal rainfed conditions (Table 3), Jute-paddy gave highest yield of 63.85 q ha^{-1} followed by jute-pea which produced a total yield of 63.85 q ha^{-1} followed by jute-pea which produced a total yield of 55.30 q ha^{-1} yr^{-1}. The lowest yield of 29.67 q ha^{-1} was obtained from jute-soyabean. The other rotations such as jute-mustard, jute-greengram/blackgram and jute-lentil produced a total yield of 34.55 q ha^{-1}, 32.10 q ha^{-1} and 31.30 q ha^{-1}, respectively.

The highest income of Rs. 11,33 oha^{-1} yr^{-1} was released from jute-paddy followed by jute-mustard (Rs. 8,293 ha^{-1} yr^{-1}) and jute-pea (Rs. 8,195 ha^{-1} yr^{-1}). The rotations jute-black gram/green gram and jute-soyabean gave net returns of Rs. 7,640 ha^{-1} yr^{-1} and Rs. 6365 ha^{-1} yr^{-1} respectively. The lowest net income of Rs. 6,130 ha^{-1} yr^{-1} was obtained from jute-lentil rotations.

Similar trend were also observed with regard to the cost-benefit ratio. The cost-benefit was highest (1.78) in jute-paddy. The cost benefit ratios were almost similar in cases of jute-green gram/black gram (1.73), jute-mustard (1.72) and jute-pea (1.70), while it was same in the case of jute-soyabean (1.67) and jute-lentil (1.67).

The success of the multiple cropping can be ensured only when the farmers are given appropriate technical know-how, adequate and timely support of inputs like fertilizers, seeds, chemicals, credit etc. besides developing the infrastructural facilities like marketing, communication and transport etc. in the villages.

TABLE 1. Jute based multiple cropping under adequate and assured irrigation conditions

Crop rotations	Variety	Sowing/ planting time	Harvesting time	Crop days
1	2	3	4	5
1. Jute	JRO 524, JRO 7835, JRO 878	Early April	Early August	109
Paddy	Ratna, IR-50, IR-36	Mid. Aug.	End Nov.	89
Wheat	UP-262, Sonalika	End Nov.	March	109
Total				**307**
2. Jute	JRO 524, JRO 7835, JRO 878	End March	Early Aug.	110
Paddy	Ratna, IR-36, IR-50	Mid Aug.	Early Nov.	89
Potato	K. Chandromukhi, K. Jyoti	Mid Nov.	February	88
Total				**287**
3. Jute	JRO 524, JRO 7835, JRO 878	End March	End July	110
Paddy	Ratna, IR-50, IR-36	Early Aug.	Early Nov.	87
Cabbage	Pusa Drum, Head Savoy	Mid Nov.	Feb.	94
Total				**291**
4. Jute	JRO 524, JRO 7835, JRO 878	End March	End July	115
Paddy	Ratna, IR-50	Early Aug.	End Oct.	86

Mustard	B-9, T-59	Early Nov.	February	94
Total				**295**
5. Jute	JRO 524, JRO 7835, JRO 878	Mid April	Early Aug.	112
Paddy	Ratna, IR-36	Mid Aug.	Early Nov.	88
Groundnut	TMV-2, J-11	End Nov.	Early April	135
Total				**355**

TABLE 1. Jute based multiple cropping under adequate and assured irrigation conditions (contd.............)

Crop rotations	Fertilizer schedule N:P:K	Spacing (cm)	Avg. yield (q ha^{-1})	Net Income (Rs. ha^{-1})	Cost benefit ratio
1	6	7	8	9	10
1. Jute	40:20:20	25x 7	28.36	5694	1.66
Paddy	80:40:40	22 x 10	37.16	4531	1.59
Wheat	100:50:50	20 x 7	27.26	2561	1.56
Total			**92.78**	**12786**	**1.56**
2. Jute	40:20:20	25 x 7	29.73	6418	1.74
Paddy	80:40:40	25 x 10	37.77	4480	1.58
Potato	120:60:60	50 x 50	155.00	7085	1.48
Total			**222.50**	**17983**	**1.58**
3. Jute	40:20:20	25x7	27.52	5855	1.72
Paddy	80:40:40	22x10	37.50	4318	1.58
Cabbage	120:60:60	75x60	254.00	8111	1.68
Total			**319.02**	**18284**	**1.67**
4. Jute	40:20:20	25 x 7	29.40	5879	1.70
Paddy	80:40:40	22x 10	36.83	4669	1.62
Mustard	60:30:30	30 x 7	10.35	2808	1.62
Total			**76.58**	**13356**	**1.62**
5. Jute	40:20:20	25 x 7	29.40	5879	1.70
Paddy	80:40:40	22 x 10	36.83	4669	1.62
Groundnut	60:30:30	30 x 7	10.35	2808	1.62
Total			**80.16**	**13328**	**1.83**

TABLE 2. Jute based multiple cropping under limited and unassured irrigation condition

Crop rotations	Variety	Sowing/ planting time	Harvesting time	Crop days
1	**2**	**3**	**4**	**5**
1. Jute	JRO 524, JRO 7835, JRO 878	End March	Early July	111
Paddy	IR-50, IR-36	Early Aug.	End Oct.	85
Mustard	B-9, T-59	Early Nov.	End Feb	90
Total				**286**
2. Jute	JRO-524, JRO 7835, JRO 878	End March	End July	109
Paddy	IR-36, IR-50	Early Aug.	Early Nov.	85
Pea	Bonavelle Arkel	Early Nov.	End Feb.	105
Total				**299**
3. Jute	JRO-524, JRO 7835, JRO 878	Early April	Early Aug.	116
Paddy	Ratna, IR-36	Early Aug.	Early Nov.	89
Tomato	Pusa, Ruby, Marglobe	Mid. Nov.	End Feb.	119
Total				**324**
4. Jute	JRO-524, JRO 7835, JRO 878	Early April	Early Aug.	110
Paddy	Ratna, IR-50, IR-36	Mid Aug.	Mid Nov.	90
Lentil	B-77	End Nov.	Mid March	110
Total				**298**

Contd......

Crop rotations	Fertilizer schedule N:P:K	Spacing (cm)	Avg. yield (q ha^{-1})	Net Income (Rs. ha^{-1})	Cost benefit ratio
1	6	7	8	9	10
1. Jute	40:20:20	25 x 7	28.79	5737	1.67
Paddy	80:40:40	22 x 10	37.15	4635	1.61
Mustard	60:30:30	30 x 7	9.37	2640	1.63
Total			**75.31**	**13012**	**1.64**
2. Jute	40:20:20	25 x 7	28.15	5993	1.77
Paddy	80:40:40	22 x 10	35.78	3797	1.65
Pea	20:40:40	30 x 50	51.90	925	2.13
Total			**115.83**	**16567**	**1.82**
3. Jute	40:20:20	25 x 7	27.85	6135	1.76
Paddy	80:40:40	22 x 10	37.35	4777	1.67
Tomato	60:30:30	75 x 0	82.30	4800	1.67
Total			**147.50**	**15713**	**1.70**
4. Jute	40:20:20	25 x 7	28.48	5918	1.70
Paddy	80:40:40	22 x 10	36.76	4357	1.56
Lentil	20:40:40	Broad-cast	8.16	2689	1.72
Total			**73.40**	**12964**	**1.65**

TABLE 3. Jute based cropping system under rainfed condition

Crop rotations	Variety	Sowing/ planting time	Harvesting time	Crop days
1	2	3	4	5
1. Jute	JRO 632, JRO 524, JRO 7835	Mid April	Mid Aug.	111
Paddy	Masuri, Pankaj	Mid Aug.	End Feb.	118
Total				**229**

2. Jute	JRO 632, JRO 524, JRO 7835	Mid April	Early Sept.	120
Pea (Gram Pea)	Bonavelle	Early Oct.	Mid Jan.	104
Total				**224**
3. Jute	JRO 632, JRO 524, JRO 7835	Mid April	End Sept.	124
Mustard	B-54, 9	Mid Oct.	End Dec.	86
Total				**210**
4. Jute	JRO 632, JRO 524, JRO 7835	Mid April	Early Sept.	123
Black gram/ Green gram	T-9, TS-16	Mid Oct.	Mid Jan	81
Total				**204**
5. Jute	JRO 632, JRO 524, JRO 7835	End April	Mid Aug.	115
Soyabean	Bregg	Mid Sept.	End Dec.	110
Total				**225**
6. Jute	JRO 632, JRO 7835	Early May	Mid Aug.	116
Lentil local		Early Nov.	End Feb.	106
Total				**224**

Contd...

Crop rotations	Fertilizer schedule N:P:K	Spacing (cm)	Avg. yield (q ha^{-1})	Net Income (Rs. ha^{-1})	Cost benefit ratio
1	**6**	**7**	**8**	**9**	**10**
1. Jute	40:20:20	25x7	26.30	2685	1.75
Paddy	80:40:40	22x10	37.55	5646	1.81
Total			**63.89**	**11331**	**1.78**
2. Jute	40:20:20	25x7	25.20	5800	1.60
Pea (Gram Pea)	20:40:00	30x10	30.10	2395	1.60
Total			**55.30**	**8195**	**1.70**
3. Jute	40:20:20	25x7	26.59	5604	1.72
Mustard	40:20:20	30x7	7.96	2689	1.70
Total			**34.55**	**8293**	**1.72**
4. Jute	40:20:20	25x7	26.00	5888	1.80
Black gram/ Green gram	20:40:00	Broad-cast	6.10	1752	1.57
Total			**32.10**	**7640**	**1.73**
5. Jute	40:20:20	25x7	24.92	5120	1.70
Soyabean	20:40:00	Broad-cast		1245	1.68
Total			1269.92	6365	**1.67**
6. Jute	40:20:20	25x7		4850	1.70
Lentil local	00:00:00	Broad-cast		1280	1.62
Total				**6130**	**1.67**

CHAPTER 5

POST HARVEST OPERATION IN JUTE

It has been observed that on an average about 86 percent of annual raw jute (jute and mesta) production is consumed by the jute industries for the manufacture of various jute products. The main exportable jute products are produced from better quality fibre. Hence, to compete in the world market, it is necessary to have better quality products and for which better quality fibre is required. Moreover, in view of the diminishing demand of packaging materials in the world market and also to the competition from synthetic fibre, jute is now being used for the manufacture of various diversified products viz. colour/printed shopping bag; decorative fabrics including wall cover, window curtains etc.; garments and other apparel cloth either from pure jute or blended with other fibre, blankets, bed linen made of jute yarns blended with wood, cotton or synthetic fibre; soft luggage and various other products. Already number of diversified units have been installed for the manufacture of diversified produce and more than 80 percent of their produce are being exported. To produce all these products, better quality fibre is required. It is, therefore, obvious that to make the jute sector a viable one, emphasis should be given to produce better quality fibre.

Jute is a bast fibre crop and the fibre is mainly obtained by extraction of the retted plant. Harvesting and post harvesting operations are the major factors governing the quality of fibre. Besides, increasing the productivity it is also necessary to improve the quality of fibre. It is, therefore, necessary to adopt proper harvesting and post harvesting operations such as retting extraction etc. in jute cultivation along with other package of practices.

Quality Character of Fibre :

Before discussing the harvesting and post harvesting operations it will be better to have an idea about the various quality character of fibre and how they are related with harvesting and post harvesting operations. The quality of fibre is mainly governed by six characters viz. i. strength ii. defects iii. root content iv. colour v. fineness vi. density . The various grades of fibre are assessed on the basis of these characters. The weightage of strength in grading is about 26%. The weightage of defects in grading is about 22%. While the weightage of root content is 33%. The weightage of colour, fineness and density are 12 %, 5 % and 2% respectively. The major quality character i.e. root content , defects and also the strength and colour depends upon the harvesting and post harvesting operations. Hence, by adopting proper harvesting and post-harvesting operations particularly retting and extraction the quality of fibre can be improved.

Post Harvesting Operations :

Retting :

Factors affecting retting : Retting is a microbiological process by which the fibres from the reeds (stem) are loosened. Bacteria and fungi act on soft tissues of the stem, which on dissolution makes it easy to separate the fibre from the wood. The retting process depends upon various factors like i. nature and volume of the water, ii. temperature, iii. pH, iv. stage of harvest of the crop and v. fertilizer use in the crop etc.

i) **Nature and volume of water** : Slow moving clear water is ideal for proper retting of jute. Bacteria eat up protein and released pectin, tannin and other gummy materials. If slow moving water is there, these impurities are removed. Soft water is good for retting and produce better quality fibre as compared to saline and hard water. In hard water lustre is affected and fibre become dazed. The ratio by volume of jute and water should be 1:20 in the retting place. Repeated charges in small ditches or tanks should be avoided without replacing the water as the fibres from the later charges leads to be of inferior quality.

ii) **Temperature :** The optimum temperature of retting water is 34°C. Higher the temperature, faster is the retting and lower the temperature, slower is the retting. At 34°C temperature the retting is completed in about 10-12 days.

i. **pH :** The optimum pH of retting water is 6.5 to 7.

ii. **Stage of harvest of the crop :** Early harvested crop retted faster than the crop harvested at late stage.

iii. **Fertilizer use in the crop :** If more fertilizer particularly nitrogenous fertilizer is used to the crop then the retting process become faster.

Method of Retting :

Retting can be done either by chemical method or biological method. In chemical method, ammonium oxalate is used and it is very costly and thereby not practiced in our country. Biological retting consist of i. stack method, ii. ribbon retting and iii. steep method. In stack method, the plant bundles are stacked and some fungal culture is used for retting. It is not practiced in our country. Ribbon retting consists of ribboning of bark from the stem by mechanical means with decorticator machine or with manual ribboner and the ribbons are then retted biologically. It is still in experimental stage in our country. Steep method of retting is practiced in our country.

Steep Method :

The jute plant bundles made after harvest are brought to retting spot. The bundles are kept standing in 50 to 60 cm deep water for 2 to 3 days which helps in retting of lower hardy portion of the bark as it requires more times than the upper portion. Few plants of dhaincha or sunhemp may be inserted in each bundles to accelerate the process of retting. The bark of these leguminous plants is soft which decompose quickly and activate the microbial process required for retting. After keeping the bundles standing for 2 to 3 days, the bundles are placed in the retting water side by side, usually in 2 to 3 days to form a sort of platform called 'Float' or 'Jack' or 'Charge'. The Jack is then covered with water hyacinth or any weeds or jute stick etc. The covering material should be such which does not release tannin or iron. The jack is then weighted down with seasoned log of wood or concrete slabs or is kept submerged at least 10 cm below the surface water by bamboo crating. Clods or banana stem should not be used as weighting materials because these materials produces dark (shyamla) fibre of low quality. Jute contains tannin and these materials contain iron combining together form ferrus tanite which is dark or black coloured. Retting can be activated by some activator like ammonium sulphate or urea or dhaincha or sunhemp. Sometimes cattle urine may be given. Incomplete submergence produce croppy fibre and also contain more bark. Proper care should be taken during retting. Retting being a microbiological process, the completion of retting is determined by inspecting few plants each day from tenth day onwards. If the fibre slips out easily from the wood on pressure from thumb and fingers, the retting is considered complete.

Extraction of Fibre :

As soon as the retting is complete the bundles are taken out of water and fibre is extracted from the stem. Two methods of extraction is usually followed, viz. i. single plant extraction method and ii. beat, break and jerk method.

Single Plant Extraction Method :

Fibre is extracted from single plant with fingers. It is a better practice of extraction of fibre and give sleak and clean fibre free from entanglement.

Beat-break-jerk Method :

Small bundles or 10 to 12 reeds (plants) are taken at a time, their stiffer root ends are beaten with a mallet to loosen the fibre. The plants or bundles is then broken at the middle and the fibre is loosened. By gripping this loosened fibre at the middle, the broken bundle is jerked in water so that the sticks slip off and fibres come out. This method often leaves broken sticks and makes fibre somewhat entangled resulting in sticky fibre.

Washing and Drying :

After extraction, the fibre is washed thoroughly on clean water in such a way that no bark or stick remains adhered to the fibres. Washing removes all the dirt and plant materials and give the fibre a bright colour. After washing, the excess water is wringed of the fibre and then the fibre is dried in mild sun over a bamboo frame. Care should be taken for proper drying of fibre. After drying the fibres are to be assorted in different grades and bundles (moraha) are to be made.

Use of Fungal Culture for quality upgradation :

Amongst the various quality character of fibre for assessing the grade, the bark (root) content has the maximum weightage. Due to improper retting the bark content may be more in fibre. At post retting stage, this bark content may be removed or reduced to improve the quality of fibre by the use of fungal culture. The National Institute of Research for jute and allied fibre Testing (ICAR) has developed the fungal culture (*Aspergillus* sp.) to treat and remove the barky portion from the jute at the post retting stage. The fungal culture, mixed with Keoline powder, is available in packets. Packets of 500 g are available and this can be used for the treatment of 100 kg of fibre.

CHAPTER 6

QUALITY ASSESSMENT AND GRADING OF JUTE

Introduction :

Jute is one of the important commercial crops in the country particularly, for the eastern and north-eastern states of the country. The price of the produces i.e. jute fibre is fixed by the Government based on its quality. But in its normal trade practices the fibre is mainly disposed of in unassorted form thereby the farmers are deprived in most of the occassions for the price commensurate to the actual quality of the fibre. Accordingly, it is necessary to make arrangement for marketing of jute fibre in graded form.

The process of sorting out fibre conforming to certain standards is commonly called grading. For efficient marketing, grading of fibre is very much essential. Because, jute is an agricultural commodity having wide variability in respect of its quality. The jute growers mostly stand to lose only because of adequate knowledge of grading is absent in them. Jute growers in general, do not know the quality of fibre they have produced. In addition to that , the growers are not thoroughly aware about the grade-wise minimum support prices announced by the Government of India every season for all the jute and mesta growing regions/states. As a result, due to their ignorance, it is found that the growers still bring jute for sale in completely unassorted form.

The purchasers, on the other hand, procure fibre simply by virtue of their personal experience and they find plenty of scope for easy manipulation in the quality of fibre and thereby the prices for gaining maximum profits out of it and the unscrupulous traders take full advantage of all situations. In the raw jute trade, grading starts after the growers have parted with the fibre.

Grading method presently in operation in our country is based on certain quality characters, which are considered at the time of their utilization for production of different materials out of the fibre. Common pattern of utilization of fibre of all grades of fibre for the production of various types of jute goods are indicated below :

Grade of fibre	Jute goods produced
W_1/TD_1	Fine yarn and specialities
W_2/TD_2	Carpet backing cloth
W_3/TD_3	Fine hessian
W_4/TD_4	Hessian wrap
W_5/TD_5	Hessian weft and sacking wrap
W_6/TD_6	Light sacking weft
W_7/TD_7	Ordinary sacking weft
W_8/TD_8	Poor sacking, cotton bagging etc.

Note : Wrap : Threads running along the length of cloth

Weft : Threads running along the width of the cloth

Hessian : For use as jute cloth only

Sacking : All types of jute bags

Thus apart, for all the products, quality wise mesta fibre is also mixed for more strength and lower cost of production of the jute goods.

At present near about 60 per cent of the total produce of fibre are grade 5 and below. The grade 5-8 are usually used for the production of sacking materials. Out of the total jute goods production near about 60 per cent is sacking. Almost the entire sacking is domestically consumed. Hardly about 2 per cent of the sacking are exported. The main exporting items are jute diversified products, carpets backing cloths, fine hessian and these materials are produced from the superior quality of fibre. If the demand of sacking is reduced in the domestic market, there will be a great set back in the demand or raw jute for the production of sacking and consequently prices of the raw jute will be reduced. Under the circumtances, it is necessary that better fibre should be produced and the fibre should be marked in graded form. The method of grading is dicussed in the foregoing chapters.

1. Old System of Grading and its Drawback :

If looked back, it was found that classification or grading of fibre was introduced in the raw jute trade on the basis of wrap yarn guarantee agreement between the Indian Jute Mills Association and the Calcutta Jute Dealers Association as early as 1935. Since then, the system of grading of jute had undergone several changes from time to time to meet their requirement of the jute of the jute mills and finally it was experienced that the entire jute produced in India was divided into several grades such as Special Top, Top, Special Middle, Middle, Bottom, B. Bottom and X. Bottom (Cross bottom). "Habi-jabi" was kept outside the above grades. The basis of grading was mainly attributed to strength, root-cuttings, colour and the place of origin.

Grading of fibre in the country, thus, was a complex task and it always differed at both the intra and inter-place or origin, season to season and from person to person. Grading of fibre was influenced much more by subjective assessment than objective measurement. Consequently, the growers neither had an idea about the quality of fibre they produced nor receive reasonable returns from their produce commnesurate with quality. This was the picture of raw jute trade in our country till mid-seventies.

In the above system of grading, assessment was done without defining the actual fibre characters needed for producing hessian or sacking etc. This created confusion to the growers as well as many others in the raw jute trade. Purchasers used to procure fibre simply by virtue of their personal experience and there was plenty of scope for easy manipulation in the quality of fibre by the trade for gaining profits out of it and the unscrupulous dealers took advantage of the situation.

2. ISI Grade specification of Raw Jute :

2.1 With a view to eliminating the above deficiencies in the quality assessment of fibre in place of the area-wise grading of fibre, formulation of a proper grading system became necessary. On the request of the Ministry of Agriculture and Agricultural Prices Commission, Govt. of India, the Indian Standard Institute presently known as the Bureau of Indian Standard (BIS) with the help of the industry, trade and growers as well as research and development institutions evolved and eight-grade system (IS:271-1975) for each of white and Tossa/ Daisee variety of jute. Fibres in this system are entirely judged on the basis of their physical properties. The new grading system was implemented in the jute grade from 1st July, 1976.

2.2 In the present system, area-wise classification was eliminated but area-wise variation in fibre quality was not totally ignored. According to quality, the entire fibre produced in the country have been standardised in eight grades. i.e. W_1, W_2, W_3, W_4, W_5, W_6, W_7 and W_8 for white jute (*C. Capsularis*) and TD_1, TD_2, TD_3, TD_4, TD_5, TD_6, TD_7 and TD_8 for Tossa (*C. Olitorious*)/Daisee jute fibre ("Habi-jabi"). However, "Hunka" has been kept outside of these grades since it's commercial value has not been considered.

2.3 The specification of each grade are based on six quality characters e.g. i) Strength; ii) Root content; iii) Defects; iv) Colour; v) Fineness and vi) Density. The present system of grading is largely scientific. System of objective measurements of the quality characters has been provided therein. But when buyers and sellers have to transact huge business within a short span of 2 to 3 months, it becomes physically impossible to carry out instrumental measurements of fibre properties methodically. So, objective assessment of "hand-and eye" methods as per the specification has been accepted for normal transactions. To reach the goal of efficient grading by "hand and eye" method, arrangements for Trainers' training for the field officers/staff at the State Level and in turn, training for the growers at the grass root level has been considered essential for proper implementation of the ISI grading system. The growers are required to know the grade of the fibre produced by them before marketing not only to receive prices commensurate with grades but also to improve upon the fibre quality.

2.4 Under the new grading system, assessment of fibre at the field level through the Hand-and-eye method may appear to be a little problemetic as the overall quality assessment may vary from person to person in respect of the eight grades provided under the specification. Even then, this could be constituted as the best since this type of scientific grading is not prevalent in any other jute/mesta growing country. After all, there is a limitation to this variation. Since, every aspect of the quality parameters is well defined, the chances of deviataion are too small. One has to remember that to accomodate every filament of jute produced in the country, it was essential to have as much as eight grades to eliminate area-wise grading. When it is talked about the farmers at the micro level, is found feel optimistic about their grading expertise since it is observed that in a particular are under the availability agro-climatic situation, the range of variation in fibre quality generally does not exceed 3 to 4 grades. Again, when it is a case with a single farm, the fibre produced therein accommodated within 2 to 3 grades. Consequently, the growers should have basic idea of grading primarily in respect of these pertinent grades. This will largely serve their purpose.

3. Implementation of the grading scheme :

3.1 For successful implementation of the grading scheme, the following steps are needed to be adopted by the concerned organization /agencies.

i) Motivational campaigns have to be organised to make the growers grade and market conscious. For dispersion of the technical know-how about grading, localized and short duration training camps, group discussion, circulation of leaf-lets, display of posters, etc. have to be arranged for the growers. This may be included in the package of practices of improved method of production of jute and mesta.

ii) Grade-wise fibre samples and corresponding prices may be displayed at the important places of the village such as Panchayat Office, market premises or agriculture office which are very often visited by the growers.

iii) Growers have to be made understood that :

a) Grading of fibre before sale is a pre-condition for jute cultivation.

b) Jute should be grown as a desired or planned crop and not as a routine or traditional crop.

c) Per hectare yield of fibre have to be raised sufficiently per hectare to minimise utilization and requirements of land, labour, inputs etc. This will ultimately bring down the per unit cost of production and make jute cultivation profitable.

d) Availability of jute sticks to be used as fuel by the growers will be more than it is available from the traditional system of cultivation.

e) Knowledge about grading will help the growers in improving the fibre quality.

4. Grading System :

4.1 The ISI specification is based on six quality parameters namely ; i) strength ; ii) root content ; iii) defects ; iv) colour ; v) density. Depending upon the importance of the quality parameters, different weightage have been given for grading of fibre. If we consider 100 marks for a good quality fibre then the weightage of different quality parameters are indicated below :

Quality parameters	Weightage (%)
1. Strength	26
2. Root content	33
3. Defects	22
4. Colour	12
5. Fineness	5
6. Density	2
Total	**100**

Before going to the grading of fibre and also details of the quality parameters it is necessory to be acquainted with certain terms required in grading system. These are elaborated hereunder :

a) **Reed :** The whole fibre system obtained from individual jute plant.

b) **Reed length :** Total length of the reed including the lower end and the tip. Though shorter reeds are generally good in quality, fibre having reed length below 150 cm have not been included in the higher grades above W8/TD8/Mesta-6 for increased handling cost in the mills.

c) **Effective reed length :** The length of the reed after the root and croppy ends have been removed. It simply means the clean portion of the fibre body. The minimum effective reed length should be 100 cm.

d) **Hunka, Moisture and natural dust :**

i) **Hunka** – The reed is covered by hard barks continuously from the lower end to almost the tip. Nowhere the fibre appears to be clean. It has not been considered as jute and as such, it has no commercial value.

ii) **Moisture** – While assessment of quality, moisture in the fibre does not affect its grade. But the weight of a jute bundle or bale having excess moisture is reduced under normal weather condition. The quality of fibre having excess moisture, very often gets deteriorated when stored for longer period.

iii) **Natural dust** : The dust or mud which sometimes get associated with the fibre during the process of its production. Retting in muddy water or placement of earthen clods/muds on the "Jack" as weighing down materials followed by improper or inadequate washing, generally develops this condition of fibre.

4.2 Quality parameters for Grading of Fibre :

The details of quality parameters, its different groups and assessment of grades are elaborated hereunder:

4.2.1 Strength : It refers to the force required to break a tuft of fibre of fixed length and of fixed wt/unit of length. To compare strength values (gm/tex), therefore, a tuft of approximate size has to be held about 5 cm apart and broken longitudinally without jerk. The whole range of strength has been divided into five classes, viz. i) Very good ; ii) Good ; iii) Fairly good ; iv) Fair average ; v) Average. Strength-wise classification are indicated hereunder :

Class	Strength values (gm/tex)		Corresponding Grade white/Tossa	Score marks allotted
	White	Tossa		
Very good	27.0 & above	29.5 & above	1	26
Good	26.9-24.0	29.4-26.5	2	22
Fairly good	23.9-21.0	26.4-23.5	3	18
Fair average	20.9-18.0	23.4-20.5	4	14
Average	17.9-15.0	20.4-17.5	5 and 6	10
Weak mixed	14.9 & below	17.4 & below	7	3 (white) 4 (tossa)

In case of mesta, strength needed for different grades has been divided into five classes :

Grade	Mesta-1	Mesta-2	Mesta-3	Mesta-4	Mesta-5
Class	Very good	Good	Fairly good	Average	Weak mixed
Score mark allotted	25	21	17	13	5

4.2.2 Root content : Hard barky region at the lower end of the root which requires additional softening for use and is normally known as 'cuttings'. The specification gives root content in terms of percentage by weight. This needs the root portion to be cut and weighed separately and the percentage is calculated on the basis of the weight of the whole reed. This can also be calculated by measuring the extent of barky portion of the root along the length without cutting it and calculating the percentage on the basis of the total reed length.

Root content allowed in different grades of jute and mesta and their corresponding score marks are indicated below :

Root content (percentage by weight) not exceeding

Grades	White	Tossa / jute	Score Daisee jute	Mesta mark allotted	Score mark allotted
1	10	5	33	12	27
2	15	10	28	20	22
3	20	15	24	30	17
4	26	20	20	40	14
5	36	26	16	50	10
6	46	35	12	–	–
7	57	42	9	–	–

Alternative method for calculation of root content is as follows:

Percentage by weight (approx.) – 2 x percentage by length :

4.2.3 Defects : Factors causing serious or partial damage to the quality of fibre are commonly known as defects. In all 12 defects have been detected in the body of fibre which have broadly been classified in two groups namely, a) major defects and b) minor defects.

a) Major defects	b) Minor defects
i) Over-retted fibre	i) Loose leaf
ii) Dazed fibre	ii) Loose sticks
iii) Centre root	iii) Specks
iv) Runner	iv) Gunny fibre
v) Knots	v) Croppy fibre and weak croppy ends
vi) Entagled sticks	
vii) Mossy fibre	

a) Major defects :

i) **Over-rooted fibre :** Fibre, which has lost its strength and brightness on decomposition as a result of retting for longer period than required.

ii) **Dazed fibre :** Fibre which is weak in strength and dull in appearance due usually to store in moist condition.

iii) **Central root :** Fibre which is more or less clean and soft at both ends but contains hard barky strips (periderm) in the middle portion. This is caused when the middle region of the stem gets exposed to sunlight after remaining submerged in water for sometime.

iv) **Runner :** Hard barky fibre running from the lower end to the middle region more less continuously. This defect can be removed by proper retting and washing.

v) **Knots :** Knots are stiff spots often barky, black or brown in colour appearing at the nodal region of the main stem or at the branch of leaf trace. Fibres near knots are not separated during retting. One of the causes may be the excreta of the apion which together with the mucilage of the plant forms a commenting material and holds the fibre strands firmly. Adoption of plant protection measure is the remedy in such cases.

vi) **Entangled sticks :** Sticks are remnants of woody part of jute plant over which fibre sheath is formed. Entangled sticks are broken sticks which are linked with fibre mass and are not easily removable.

vii) **Mossy fibre :** It is a type of vegetation which sometimes gets attached to the jute plant during flood conditions ; portions of it may remain on the jute fibre even after retting and washing. It can be separated by hand or by combining.

b) Minor defects :

i) **Loose leaf :** Dark grey leafy or paper like substance (remnants of loosened skin of the plant) appearing on the strand. Loose leaves are those that lie loosely in the fibre and are easily removable.

ii) **Loose sticks :** Loose sticks are broken sticks easily removable by shaking.

iii) **Gunny fibre :** Fibres held together by undissolved pectinaus matter. Proper retting and washing may remove these defects.

iv) **Speaks :** These are darky spots, normally oval shaped, dark grey to brown in colour, not larger than 1.5 cm. These spots may be due to punctures by female apion, an insect or due to infection caused by the fungi, which damage the cortical tissues. Sometimes improper washing of fibres leave barky spots. Fibres in the speaky region can be separated with some effort but that remain as weak spots. Adequate plant protection measures and proper retting and washing may help in removing this defect.

v) **Weak croppy end :** Fibre over a length of about 30 cm at the top end which has become unusually weak.

vi) **Croppy fibre :** Fibre with top ends rough and hard (but not barky) caused by careless retting. Defects rejected and allowed in different grades and corresponding score marks allotted :

JUTE

Grade	Defects rejected/allowed	Score marks
W_1/TD_1	Free from all major and minor defects	22
W_2/TD_2	Free from all major and minor defects	22
W_3/TD_3	Free from all major and minor defects except some loose leaf and a few speck	18
W_4/TD_4	Free from all major defects and substantially free from loose sticks and specks	14
W_5/TD_5	Free from all major defects	10
W_6/TD_6	Free from centre root, over retted and dazed fibre and reasonably free from entangled sticks	4
W_7/TD_7	All major and minor defects are allowed	0

MESTA

Grade	Defects rejected/allowed	Score marks
Mesta-1	Free from major and minor defects except brown leaves and appreciably free from specks	25
Mesta-2	Free from major defects and loose sticks ; substantially free from knots and gummy and croppy fibres	21
Mesta-3	Free from major defects except knots	17
Mesta-4	Free from centre root, dazed and over retted fibre and reasonably free from runners and entangled sticks	13
Mesta-5	Reasonably free from runners	5
Mesta-6	All major and minor defects are allowed	0

4.2.4 Colour : The word colour indicates the degree of cleanliness or whiteness or brightness of fibre. It also acts as a subjective indicator for strength as well as gives clear idea about the retting situation and retting method adopted. By arranging natural colours of fibre, designed jute cloth are also prepared. The whole range of fibre colour has been divided into five classes. They are :

JUTE : Description of fibre colours

Class	White jute	Tossa jute	Degree of brightness	Corres-ponding grade	Score marks allotted
i) Very good	Light creamy to white	Golden reddish white	65 & more	1	12
ii) Good	Creamy pink to brownish white	Reddish to brownish white	64-56	2	9

ii) Fairly good	Brownish to reddish white to light grey	Reddish or brownish to some light grey	55-36	3	7
iv) Fair avg.	Brownish to light grey	Light grey to copper colour	35-31	4	4
v) Avg.	Grey to dark grey	Grey to dark grey	30 or below	5	3

MESTA

Class	Colour description	Corresponding grade	Score mark allotted
Good	Creamy to whitish	Mesta-1	15
Fair	Light grey	Mesta-2	11
Average	Greyish to dark grey	Mesta-3	7

4.2.5 Fineness : Width or diameter of filaments as judged by closely observing the filaments after the fibre strands are opened by the manipulation of fingers. Very fine, fine and well separated fibres have been found to be significant in higher grades. In lower grades, fineness has not been considered. The whole range of fineness has been divided in the following way.

JUTE

Class	Very fine	Fine	Fibres well
Range of values (tex) White jute	1.8 & below	1.9-2.4	2.5 & above
Tossa jute	2.7 & below	2.8-3.4	3.5 & above
Corresponding grades	1	2	3 and 4
Score mark allotted	5	2	1

In case of mesta, the fineness has been judged as finer with fibre well segregated and fine with fibre well segregated in Mesta-1 and Mesta-2 grades respectively. Coarse fibre has been considered for Mesta-3 grade.

MESTA

Grade	Mesta-1	Mesta-2	Mesta-3
Class	Finer with fibre reeds well segregated	Fine with fibre reeds well segregated	
Score mark allotted	5	3	1

4.2.6 Density : Weight per unit volume of fibre including air space. It may be the heaviness of some fibre reeds, held within the grip and raised up and down. Only two classes, i.e. i) Heavy bodied and ii) Medium bodied fibres have been considered.

JUTE

Class	Heavy bodied	Medium bodied
Values (gm/cc) White and Tossa jute	0.4 and above	0.44 and below
Corresponding grade	1 and 2	3 and 4
Score mark allotted	2	1

MESTA

Class	Very heavy with thinner reeds	Heavy with broader thicker reeds	Medium bodied
Corresponding grade	Mesta-1	Mesta-2	Mesta-3
Score mark allotted	3	3	1

Requirements for each grade of White and Tossa/Daisee jute as well as Mesta as per BIS grade specification (IS:271-1975) and (IS: 9846-1981) are placed below in a summarized form for easy understanding.

5. *Overall grade assessment as per BIS specification for Jute (IS : 271-1975) :*

Grade	Characters	Score
W_1/TD_1	Very good strength and colour; very fine heavy bodied fibre; free from major and minor defects; maximum root content allowed W_1 : 10% and TD_1 : 5% . Total score marks allotted.	100
W_2/TD_2	Good strength and colour ; fine heavy bodied fibre ; free from major and minor defects ; maximum root content allowable W_2 : 15% and TD_2 : 10% ; total score marks allotted .	85
W_3/TD_3	Fairy good strength and colour ; well separated medium bodied fibre ; free from major and minor defects except some loose leaf and a few species ; maximum root content allowed - W_3: 20% and TD_3 : 15%. Total score marks allotted.	69
W_4/TD_4	Fair average strength and colour; well separated medium bodied fibre; free from major defects and substantially free from spikes and loose sticks ; maximum root content allowed : W_4 : 26% and TD_4 : 20%. Total score marks allowed.	54
W_5/TD_5	Average strength and colour ; free from major defects ; maximum root content allowed : W_5 : 36% and TD_5 : 26%. Total score marks allowed.	39
W_6/TD_6	Average strength ; free from centre root, over retted and dazed fibre and reasonably free from entangled sticks ; maximum root content allowed : W_6: 46% and TD_6 : 35. Total score marks allowed.	26
W_7/TD_7	Weak mixed fibre ; maximum root content allowed - W_7: 57% and TD_7 – 42%.	12/13
W_8/TD_8	Entangled or any other jute not suitable for any of the above grades but of commercial value.	0

6. Requirement of fibre characteristics for white and tossa jute and mesta :

a) The minimum reed length should be 150 cm. or the effective reed length should not be less than 100 cm. except for W8/TD8 and mesta-6.

b) Fibre should be free from Hunka, mud and other foreign materials and in dry storable condition.

c) Natural dust may be allowed in grades W_5/TD_5 to W_8/TD_8 and Mesta-4 to Mesta-6 with proportionate discounts.

d) Root content will include hard barky croppy ends.

e) A parcel of fibre which would not score full marks for a particular grade shall still be considered for that grade with suitable discount to be settled between the buyer and the seller, provided its score is not less by 50(or more) per cent of the difference between the maximum score for that and the next lower grade. When the score is less by 50 (or more) per cent of the difference, the buyer will have the option to reject or settle with a suitable discount.

7. Method of assessment :

In the field or in the markets when the volume of fibre is large, for quick assessment of the quality characters a grader may have his observations on the basis of so called "Hand-and-eye" method. While 'hand' may feel the hardness and sense strength and heaviness; 'eye' will judge colour, root content, defects and fineness of fibre. However, a grader should follow the scoring system, already described above, for his guidance.

□□□□□
□□□

CHAPTER 7

COST OF CULTIVATION OF JUTE

It is established fact that no producer will prefer to produce a thing or commodity, which, in turn, will not generate worth while return or economic and social benefit. In this context, the cost of production assumes to be the single most determining factor to justify the continuation of production, particularly in an arena of global economic competetiveness. Supply of any commodity is bound to be even or uninterrupted if it proves to be gainful and it is bound to be uncertain and uninterrupted, once the product is reckoned as uneconomic and unviable. Economic criterion suggests that a product should be as much cost effective as possible and its marketing spectrum should simulteneously be adequately efficient to cater the need and serve the interest of producers, sellers ans consumers alike. It may be mentioned that for increased production and sustaining the level, the growers should receive remunerative price for their produce in relation to their cost of production, otherwise there is a risk of shift away from jute to some other more profitable crops as usually happens in years following a crash in raw jute prices.

Cost study is a very complicated one. If the actual cost of cultivation of jute is to be calculated then so many related cost has to be considered starting from higher of human labour, value of owned bullock labour, value of higher bullock labour value of owned and higher machinery, value of fertilizers, value of seeds, value of insecticides and pesticides, irrigation charges, land revenue, cesses and other taxes, depreciation of farm machinery, implements and farm buildings, interest on working capital, miscellaneous charges etc.

If a rough estimate of cost of cultivation of jute is made, the cost may be written as follows.

Break up of cost of cultivation per hectare (Rs.)

Item no.	Item	Cost
1.	**Operational cost**	9236.31
2.	**Human labour**	
	i) Casual	3350.51
	ii) Attached	663.96
	iii) Family	2450.44
	iv) Total	6,464.91
3.	**Bullock labour**	
	i) Hired	234.00
	ii) Owned	738.99
	iii) Total	972.99
4.	**Machine labour**	
	i) Hired	1.43
	ii) Owned	7.88
	iii) Total	9.31
5.	**Seed**	391.66
6.	**Fertilizers and Manure**	
	i) Fertilizers	554.56
	ii) Manures	276.53
	iii) Total	831.09
7.	**Insecticides**	142.51
8.	**Irrigation charge**	205.63
9.	**Fixed costs**	
	i) Rental value	5239.94
	ii) Rent for based land	20.59
	iii) Land revenue, cesses, taxes	14.65
	iv) Interest on fixed capital	409.73
	v) Depreciation on implements and buildings	160.97
	vi) Total fixed cost	15082.19

Jute is a labour intensive crop. Human labour comprises 69% of operational costs. Jute is being cultivated by using family labourers. About 50% labour costs represent family labour. Family-bullock use is high in jute cultivation. 90% bullock power comes from home. Farm mechanisation in jute cultivation is almost negligible. Output price indicates instability of price which leads to instability of income from jute. Low level of capital is being used for growing jute. 80% of the operating costs represent human labour and animal labour. Level of insecticides and fungicides use is low. Management of inputs of jute is poor.

During calculation of cost of cultivation of jute instability of prices, fluctuation of yields, market risks, anxiety of production and marketing etc. should be included carefully. Cost as well as yield of produce is an instable component.

CHAPTER 8

SEED PRODUCTION TECHNOLOGY OF JUTE

Jute being a best fibre crop, both seed and fibre of good quality cannot be obtained from the same plant and the quality of fibre deteriorates if the crop is left standing until the seed maturity. Some farmers keep a small portion of the crop at a corner of their fields to produce seed and the rest is harvested for fibre. It has been estimated that the average area retained for the seed production is about 4.1 percent of the total average under jute. However, seeds produced in West Bengal are not sufficient to meet the demand of jute growers of the state. As a result the major part of supply of seeds for West Bengal is met from the other states like Maharashtra, Gujrat and adjoing dry areas of Bihar and Orissa. If effort is made good quality seed can be produced in potential seed growing areas in West Bengal. This includes the districts of Purulia, Bankura, Burdwan and Midnapore.

The following package of practices should be taken into consideration for successful production of jute seeds.

Seed :

Seed must be procured from authentic source like Central Research Institute for Jute and Allied Fibres, National Seed Corporation, Seed Multiplication Farm etc. Standard varieties suitable for different agro-climatic regions should be used to raise a seed crop. Germination percentage of seed should be 90 or above. Before sowing the seed material from a particular lot should be tested. A good seed sample should have 98% geminability at 33°C temperature. Capsularis takes about 24 hrs. and olitorius requires 48 hrs. for germination in petridish.

Selection of Land :

For raising good quality seed, a land where same kind of crop was not grown in the previous year, should be selected. The land should be fertile, well drained and neither acidic nor alkaline. The pH of the soil should be around 6.4. If the soil is very acidic, it should be corrected by application of dolomite at the rate of 1 ton per hectare at least 30 days before sowing. It should be free from obnoxious weeds and soil borne diseases like stem-rot.

Isolation Distance :

Jute is self-pollinated crop but cross pollination does occur to some extent . It has been estimated that natural cross pollination in Capsularis usually ranges from 2 to 5% and as high as 12% in olitorius jute. For foundation seed production the seed field should be isolated by 50 m whereas in case of certified seed production the isolation distance is 30 m. Isolation distance of only 3 m should be provided between two species of jute.

Preparation of Land :

The land should be prepared well by deep ploughing and cross harrowing followed by laddering to break the clods until a fine seed bed with good tilth is obtained as fine tilth is necessary to ensure good germination. Weeds and stubbles of previous crop should be removed at the time of final bed preparation. The beds should be raised 4-5 cm. high with shallow drainage channel in all the sides to avoid possible water stagnation.

Seed Treatment :

Prior to sowing, the seed material should be treated with fungicides against seed borne pathogens at the rate of 5 g Aretan or 2 g Bavistin kg^{-1} of jute seed. Seed with such chemicals are to be mixed thoroughly in a rotary drum for at least 10 minutes.

i) **Seed rate** : Seed rate for Capsularis is 7 to 10 kg ha^{-1}, whereas, in case of olitorius 5-7 kg ha^{-1}. The optimum seed rate with good quality seeds ensure an even emergence of seedlings and later seedlings are thinned out between the plants maintaining a spacing of 10-15 cm. At the time of thinning care should be taken to remove weeds with a hand hoe.

ii) **Time of sowing :** For seed production sowing may be delayed to take the advantage of good moisture from delayed monsoon in dry areas and also delayed harvesting which is good for a seed crop as the weather expected to remain dry at the time of harvest. But the best time for seed production is in between April and May.

iii) **Method of sowing :** Line sowing with the help of a seed drill in a spacing of 30 cm. for Capsularis and 25 cm. in case of Olitorius in between rows facilitates intercultural operations like rouging and inspection of seed crops.

Seed may also be sown behind the plough maintaining a space as in row crop. During broadcasting half of the total seed is to be sown from North to South and the East to West direction. After broadcasting a light laddering is given for proper compaction of seeds.

Spacing and Plant Population :

For a seed crop profuse branching is encouraged as the number of fruits and quantity of seeds obtained would be much more. This is quite in contrast with the fibre crop where branching is considered to be a bad quality. Thus to make more branching plant population is maintained at a lower order by producing wider spacing and a low seed rate. Row to row spacing is maintained 25-30 cm and plant to plant 10-15 cm.

Manuring :

Seed crop require manuring just to maintain optimum level of soil fertility that is not less and not more. A more fertile soil will have a tendency to make the plants over vegetative which is not desirable for a seed crop. A less amount fertilization may not be able to make the branches fully developed and flowers, fruits and seeds properly formed. Thus the dose of nitrogen is drastically reduced and those phosphorus and potassium moderately reduced. The recommended dose of NPK fertilizers for a seed crop is 15:15:15 kg ha^{-1}. Entire quantity of phosphatic and potassic fertilizers and 50 percent of nitrogen should be applied at the time of sowing and the rest nitrogen should be top dressed one month after sowing and under adequate soil moisture condition.

Irrigation :

Prior to sowing irrigation may be applied for land preparation and subsequent 2-3 irrigations may be required before the onset of monsoon.

Intercultural operation :

The weed infestation of jute is maximum upto sixth week of crop age and frequent wheel hoeing at this stage is necessary for improving tilth, aeration and to preserve moisture. This operation is done 2-3 weeks after sowing and thereafter weekly or bi-weekly intervals. Besides, two hand weeding as well as thinning of seedlings to a space of 10-15 cm. should be completed within five weeks of crop age.

i) **Rouging :** Rouging for off types, diseased and pest infected plants should be carried out at 30-40 days crop at flowering and capsule formation stages.

ii) **Pruning :** Pruning of apical bud six weeks after sowing indicates profuse branching and leads to a better seed yield. Pruning should be done at a place on the stem 5-7 cm. from the top. Soil should contain sufficient nutrients to support the quick growth of all branches.

iii) **Cutting :** Stem cutting of jute plants strike roots when planted in well prepared nurseries. One who likes to produce his own seed may decide to harvest a part of the jute crop well ahead of flowering at 90-100 days when the crop attain 75-80 percent of growth. After harvest of jute the apical portion is cut off and plant them first in well prepared nursery and then in the main field and get quality seed and remaining portion of the harvested stem would produce good quality fibre although yield may be less.

iv) **Plant protection measure :** To produce disease free seeds it is most essential that necessary protection measures taken right from the beginning to control the incidence of diseases and pests.

In jute stem-rot and root-rot diseases appear mostly. To control them bavistin at 0.1 percent may be sprayed 2-3 times.

For controlling pests like jute Semilooper and Bihar hairy caterpiller endosulfan or phospapalone at 0.075 percent is to be sprayed at weakly or bi-weakly intervals.

Harvesting and Threshing of Seed :

Harvesting is done at full maturity when seed pods become brown. In C. capsularis good quality seed is obtained by harvesting when stems and pods are not dead ripe. Harvesting is done by a sharp sickle during the morning hours and the plant bundles are kept on the threshing floor for drying.

Threshing is done by beating the pods with the help of a wooden stick and then the seeds are winnowed and sun dried to reach the moisture content below 9 percent and kept for storage. Precaution should be taken to avoid contact of seeds with soil as it spoils the appearance and may increase moisture as well.

Storage :

Jute seed losses viability rapidly if it is not stored properly. Hence, after winnowing, seeds should be properly dried till the moisture content of the seeds reduces to 9 percent. Immediately after drying the seeds are allowed to cool and stroing is done on the same day in good quality double gunny bags or gunny bags lined with alkathine cloth. Earthern jars may also serve as container provided it is plastered with mud and cowdung.

Small samples may be stored in glass bottles with screw cap. It may be mentioned that jute seed losses viability when there is prolonged contact with damp air. Hence the godown for jute seed should be dry and jute seed bags are not to the kept in contact with floor.

Yield :

Seed yield of C. Capsularis is 3-4 q ha^{-1} and in case of *C. Olitorius* 1-15 q ha^{-1}. However, on an average seed yield of 4-5 q ha^{-1} and 3-4 q ha^{-1} for *C. Capsularis* and *C. Olitorius* respectively can be obtained if proper care is taken right from sowing of seeds to harvest.

❑❑❑❑❑
❑❑❑

CHAPTER 9

MARKETING OF JUTE

Marketing in a broad sense and in common parlance means buying and selling, involving exchange of goods and services between a person who possesses then and person who requires them. Marketing has been defined as "those business activities involved in the flow of goods and services from production to consumption". An agricultural marketing includes all processes which are under taken to move the agricultural commodities from fields to factories and from producers to consumers. In the following paragraph, the discussion is confined to the marketing of raw jute.

Jute popularly known as "the golden fibre" occupies a pivotal place in Indian economy, particularly of the eastern region. This is apparent from the tremendous socio-economic importance of the crop. Broadly speaking, about 40 lakh farm families of the jute growing states of West Bengal, Bihar, Assam, Orissa, Uttar Pradesh, Tripura and Meghalaya are engaged in the cultivation of jute Mesta, another bast fibre crop is also grown in these states, but it is exclusively grown in Andhra Pradesh. The jute industry which is the ultimate consumer of raw jute provides employment to nearly 2.5 lakh workers besides providing job opportunity to another 20 lakh people in the secondary and tertiary sector of the jute industry. Jute and mesta are used as raw material for the jute industry which is notably export-oriented. Currently the jute industry is earning about Rs. 300 crores annually as foreign exchange through export of jute in about 120 countries and an all out effort is being made to raise the level further to about Rs. 1000 cores.

Giving due cognigence to the importance of this crop, the Govt. of India has all through gone for a systematic drive in all the plan periods to boost the production and productivity of the crop along with a simultaneous improvement in fibre quality. As an outcome of this efforts, production is largely established to meet the total requirements of jute industry calculated at about 80 lakh bales. Supply is comfortable as compared to demand and, basically there is no need to import raw jute.

Marketing scenario in raw jute in India is by and large, middleman-oriented and different market functionaries are involved in its trade in different categories of markets. It may be worthwhile to begin with the tiers of jute markets.

i) **Tiers of jute Markets :** In the marketing of raw jute, the movement of fibre from the producer to the consumer (jute mill) takes place in three distinct stages through a chain of middleman; the first stage is from the villages to the primary assembling markets ; the second stage is from the rural areas to the secondary markets, viz. baling centres and the third stage is from the baling centres to the terminal market of Kolkata or to the concerned jute mills of other states.

ii) **Market functionaries :** In the channel of distributions, different market functionaries are found to operate in the jute trade between the growers and jute mills.

Farias, paikars and beparis who move about the villages form the first link in the chain of intermidiaries. A large portion of the marketable surplus of jute is disposed of by the growers at their door steps. Aratdars and balers also operate in the villages and primary markets on commission basis. The primary market, thus consists of 'hats' and 'shamdies' usually setting once or twice a week, where the jute growers dispose of their produce to the faries and beparis as also to the purchasers on behalf of balers operating in the secondary markets.

The secondary markets is distinguished from primary market by interpreting it as a place where baling is done. These markets are located at places where communications are good as well as large and regular supply of commodity is assured.

In the secondary markets 'kutcha' balers are the most important and powerful group of functionaries responsible for assembling, processing, storing, financing and transacting of jute fibres.

In the terminal market of Kolkata, the functionaries are Aratdar, Dalals, Pucca balers and jute Mills. Obviously the secondary markets form the most important link between producers and ultimate consumers. In fact, prices in the villages and primary markets are affected to a great extent by the prices and marketing practices prevailing to the baling centres.

iii) **Sale of jute at different levels :** In a study, it was found that on an average 37 percent of the marketable surplus is sold by the growers at villages, 56 percent at primary markets and hats and only 7 percent at secondary markets.

Commencement of Sale and Peak Period of Marketing :

Assam and sub-Himalayan West Bengal are usually first to harvest the white jute crop in July followed by Bihar, Orissa and Uttar Pradesh. Harvesting of tossa jute also commence first in Assam and Sub-Himalayan West Bengal and Orissa. Mesta crop is harvested last during September- October through out the country.By and large, marketing of jute commences in Assam, sub-Himalayan West Bengal and eastern part of Purnea district in Bihar in July and reaches its peak in October – November throughout the jute growing tract of the country baring Orissa where due to late sowing marketing commences in September and the largest sale is recorded in December. In Tripura, the marketing starts in August and reaches its peak in November-December. In Andhra Pradesh, the marketing of fibre starts in mid September and reaches the peak in November-December. The pattern of rainfall in a particular year, however, causes slight deviation.

Trend of Fibre Arrivals :

Seventy five percent of the total fibre is marketed in the primary markets during September-December period and recording its peak in October which incidentally happens to be as important festival month. Arrival of fibre starts declining thereafter. Transactions are appreciably low and virtually nominal during March-June period. In the secondary markets, over 65 percent fibre is transacted during September-December period. Transactions at the terminal level are found largely even and steady during different months.

Price situation of Raw Jute :

Pricing of raw jute is directly linked with the supply and demand of the commodity as well as the expected total availability of the produce during the season. The peculiarity of jute marketing is that commodities like food grains have a multidirectional movement from farm to the consuming centres while in the case of jute there are large number of production or supply points scattered over the jute/mesta growing states as compared to few consumption points or terminal markets located in and around Kolkata or state capitals. Again, the entire crop of raw jute equivalent to twelve months consumption of mills being taken out of the field in 3-4 months period and marketed by the growers in only a period of 4-5 months. Consequently, the pressure of the supply at a given point of time is much more than the demand which can not be taken care of by the consumers alone who would normally like to have a regulated and even supply for the mills throughout the season. Because of an imbalance between the supply and demand, fibre prices are invariably quoted low during September-November period when supply outstrips demand. The highest prices of the commodity is recorded during the last quarter of the season when arrival tapers off rapidly and fibres no longer held by the growers but by the trade. The jute cultivators generally belong to small and marginal category and they are forced by the economic pressure to dispose of their produce within the initial few months of the season.

Price spread amongst Tiers of Markets :

Price spread or the market margin retained by the market functionaries varies widely amongst different tiers of markets. Price risk arising out of time lag and the associated marketing expenses prompt the marketing agencies to quote prices in such a way as to leave them with as much price differential as possible between the aggregate purchase price vis-à-vis the aggregate sale price for maximization of profit. In a study it was found that prices paid to the growers varied between 69 percent and 83 percent of the prices paid by the ultimate consumers depending upon the distance of the centres in relation to the ultimate point of sale.

Production Trend :

The situation of raw jute production over last two decades show that the volume of production in India revealed a upward drift from about 910 thousand MT in mid seventies to 1442 thousand MT in mid eighties accounting 58.6 percent growth. While the growth of output over eighties to nineties is accounted barely 0.8 percent indicating a state of stagnation. Area under jute cultivation is also noticed to have similar trend. This trend of results is obtained throughout the country in which states jute are cultivated. It is also true that with the years, the area under jute cultivation decreases and the area is covered by cereal / oilseed/pulse crop. A substitution of fibre crop by the food crops would generate negative impact on income and employment due to the fact that crop used to have relatively less in built capacity to generate income and employment from various stages of value addition activities through which it moves till its cultivation to end product.

Markets of Jute Products :

India's own produces and import of jute fibre are processed within the country for the home and foreign markets. Jute and jute products enjoyed a monopoly market advantage in home and abroad, which was challenged with emergence of synthetic fibre in early sixties of the last century. The principal traditional jute products namely, jute twine, yarn, hessian , sacking, carpet backing, clothes etc. would have been used as packing materials for food grain, sugar, cement, fertilizer, salt, vegetables, cotton and also postal canvass – in home and abroad. With the advent of indigenous synthetic packaging units in late seventies the above noted traditional jute products experienced a serious set back. Thus , with the emergence of substitutes the market prospect of jute and jute products in the home and abroad needs to be looked into taking into account a range of its synthetic substitutes of diversified nature. In the present competitive market regimes demand creation for jute products can effectively be understood if one addresses the changes in consumers' taste and preferences with the change of their objective conditions. In the present situation it needs to be viewed from the side of consumers' preference rather than producers' and manufacturers' demand for traditional jute products meant for packaging. This implies a need of transformation in the use of jute fibre from packaging to textile and non-textile areas like travel goods, composites, paper and jeojute.

Average Area ('000 hectares) and Production ('000 tonnes) of jute in major jute producing states of India for the last three decades

State	Decades					
	70's		80's		90's	
	Area	Production	Area	Production	Area	Production
Assam	99.6	118.4	108.1	108.6	92.5	156.1
Bihar	124.8	120.3	163.4	217.6	114.1	161.6
Orissa	44.3	66.9	39.3	62.6	22.8	42.1
West Bengal	418.0	577.5	577.3	958.5	529.4	1084.6
India	706.7	909.2	882.1	1442.3	765.2	1454.5

Marketing Inefficiencies :

Jute market is, by and large, middlemen-oriented. Forces of unregulated trading brought in different market functionaries at various stages of development of jute markets and in the process, the marketing structure of jute has developed many functional weaknesses and frictional defects arising mainly out of market imperfections, some of which are discussed below.

i) The jute growers, is by nature, a simple producer, unaware of the intricacies of modern marketing. He is usually resourceless and has to sink all his capital in the land. He needs finance for raising the crop, for harvesting, for marketing, transporting etc. Even for securing loan, he has no adequate security to offer and he is always in financial distress. He has no bargaining strength. Jute markets on the other hand, are controlled by the traders who are very much resourceful and organised while the growers are scattered and highly deorganised.

ii) Development of the transport and communication system is vital for efficient marketing. Imperfections in marketing structure as also spatial disparities in prices have, in many cases, created problems in transportation. Transport system in the rural areas are not yet properly developed and this prompt the jute growers to despair of their produce to Farias and Beparis in the villages who have the resources to effect transport.

iii) Proper storage facilities are essential for efficient marketing. They help in tiding over the deficits of particular periods and areas and thus act as an equalizer of prices both regionally and over period of time. Most of the jute growers have neither storage facilities nor the capabilities to hold fibre. Consequently, they take part with most of their marketable surplus within a few months of the harvesting of the crop.

iv) The facilities for cheap credit are very much limited at the grower level. Indebtedness amongst the growers are widespread. The existing institutional credit facilities are not only inadequate but also complicated. Consequently, the growers are completed to the loan from the unscrupulous village money-lenders and jute traders who either charge an exorbitant rate of interest or make advances under the 'Dadan' system is contractually obligatory to sell fibre to them, generally well below the prevailing market rates.

v) Dissemination of proper market intelligence is an essential element in market transparency for ensuring fair and remunerative price to the farmers. But there is no such independent source of market intelligence for growers who have mostly to rely upon dealers for price quotations either in the up-country secondary markets or in the terminal market. The traders take undue advantage of the growers' ignorance of the ruling prices.

vi) Grading of jute/mesta is essential to help the farmers to receive fibre prices commensurate with grade or quality. But most of the growers are unaware of the ISI grade specification and they dispose of their produce in an unasserted or 'garsat' form. Consequently, they fail to receive just prices for their fibres in terms of quality.

vii) A large number of arbitrary market allowances and deductions both in cash and kind exist in the jute trade which are eventually borne by the sellers. Such allowances and deductions refer to dhalta ; pelladari, dalali etc.

viii) Requisite infrastructural facilities are essential for orderly and efficient marketing of agricultural produce. But most of the basic infrastructural facilities are found absent in the primary stage and, as a result, the jute growers who come to the market with their produce are compelled to make immediate sale even under unfavourable terms.

ix) Cooperative marketing of jute is known to be an ideal way to safeguard the interest of the growers which can eliminate the fundamental weakness in the structure of markets and improve upon the bargaining power of the growers. The cooperatives can build up a strong sellers' organisation to match the powerful buyers' organisation. But these has not been the required development on this front.

Price Policy :

The price of the Govt. is to provide proper price to the farmers for their produce. In order to remove the structural weakness of jute marketing, the Govt. of India created the jute Corporation of India (JCI), a public sector undertaking in 1971. This organisation is committed to provide at least the minimum support price to the raw jute growers.

CHAPTER 10 PROBABLE TECHNOLOGIES FOR MANUFACTURING JUTE DIVERSIFIED PRODUCTS

Jute is an important commercial crop and provides sustenance to a large section of rural households mostly in the eastern part of India. Jute is a natural, biodegradable, non-toxic and eco-friendly fibre. So long jute was mainly used as a low value industrial fibre for manufacturing coarse fabrics for packaging. With the change in global scenario and growing consciousness about eco-friendliness newer diversified value added products from jute are emerging and are being marketed. Traditional practices of handloom wearing based on jute may play an increasingly important role in income and employment generation in the coming decades. Among the major constraints hindering the speedy growth of jute industries in rural and semi-urban areas are non-availability of appropriate technologies, poor coordination between research organisations and enterpreneurs and sometimes uncertainty in raw material supply.

Products suitable for Rural Sector :

Products ranging from vegetable and grain packaging to have textile, soft luggage, shoe uppers and travel aceessories can be produced. The products may be improved in appearance and can be given finishing treatments to make them resistant to water, mildew, rot and flame. Markets of there products is expanding. Following is an itemwise description of the products.

A. All jute fabrics :

Inputs : Commercial jute yarn (6 lb, 7.5 lb and 9.5 lb)

a) Light weight fabrics (110-220 gm m^{-1})

i) **Scrim :**

Wrap : 8 lb, 5 ends inch^{-1}

Weft : 8 lb, 4 picks inch^{-1}

Approximate area density : 110 gm m^2

Use : Fruit bag, lamination and lining in furniture industry.

ii) **Light Hessian :**

Wrap : 8 lb, 9.5 ends inch^{-1}

Weft : 8 lb, 8 picks inch^{-1}

Approximate Area density : 213 gm m^2

Use : Potato bags, Onion bags etc.

b) Medium weight fabrics (300-400 gm m^{-1})

Wrap : 8 lb, 12 ends inch^{-1}

Weft : 9.5 lb, 12 picks inch^{-1}.

Approximate Area density : 305 gm m^2

Use : Fabrication of shopping bags, soft luggage, furnishing fabrics, cereal bags etc.

B. Jute-Cotton Union Fabrics :

Inputs : Commercial jute and Cotton yarns.

Yarns may be bleached and dyed prior to the wearing. Sizing of wrap is to be done when necessary. Generally, the cotton yarns is used as the wrap. However, jute can also be used as wrap when the production is likely to be lower than when cotton is used. Following is a representative list of products.

Sl. No.	Products	Fabric weight (gm m^{-1})	Jute content (%)
1.	Fabrics for soft luggage	410-425	70
2.	Furnishing fabrics	300-400	60-65
3.	Mats	450-1200	70-95

Unit Operation required to convert Yarns into Fabrics / Products :

The process of converting yarns through fabrics to value added products would be the following broad operations:

i. Fabric designing, yarn (hank) dying, hank sizing and wrap preparation includes drawing and deating.
ii. Weaving including weft preparation.
iii. Chemical processing of fabrics and yarns.
iv. Lamination of fabrics for soft luggage.
v. Fabrication of bags and soft luggage.

These activities form the basis of industrial projects for rural sector.

Possible projects for rural sector :

Manufacturing of Jute and Jute-cotton-linen Fabrics

Economic Aspect :

i. Land and building : Covered area 100 M^2
ii. Machinery and Equipment : Rs. 1.20 Lakh
iii. Labour (9) : Rs. 15,000 per month
iv. Working capital : Rs. 1.46 lakh(for 3 months)
v. Production capacity : 30,000 meters annum^{-1} (10% all jute and 90% Jute-cotton)
vi. Cost of production (Rs 6.27 lakh annum^{-1} including depreciation and bank interest)
vii. Break even point : 47%

Chemical Processing of Jute based Textile (Yarns and Fabrics) :

Chemical processing units may be set up to convert grey fabrics into aesthetically dyed or printed goods using 100% jute or blends/unions of jute with cotton, wool, viscore rayon, acrylic, polypropylene etc. in different proportion. Chemical processing would be carried out in branches in a hank bleaching/dyeing unit for yarns and in jigger for fabrics. Depending upon the type of product desired the materials would be subjected to unit operations like, scouring, bleaching, dehairing, dyeing, printing and finishing. Washing and hydroextraction would be employed when necessary.

Economic aspect :

i.	Land and building	: land area 100 sq m costing Rs. 30,000 and building area 50 sq m amounting Rs 1.00 lakhs.
ii.	Machinery and equipment	: Rs. 0.8 lakhs
iii.	Labour and staff	: Rs. 10,000 per month
iv.	Working capital (for 3 months)	: Rs. 1.8 lakhs
v.	Cost of production per annum	: Rs. 8.39 lakhs
vi.	Production capacity (p.a.)	: Bleaching only 30 tonnes. Bleaching and dyeing 42 tonnes.
vii.	Break even point	: 51.99%

Manufacturing of Soft Luggage using Jute and Jute-blended Fabrics :

Jute based soft luggage such as shopping bags, school bags, folio bags, brief cases etc. have received wide consumer acceptance in India and abroad. Bags and soft luggage prepared out of jute based fabrics are eco-friendly and hygienic besides being competitive to cost when compare to synthetic products. Laminated jute based fabrics processed from the market are generally used for manufacturing such products. The fabrics are generally bleached and dyed before lamination. These are cut to pieces of desired sizes employing templates. The pieces with necessary accessories are then stitched together with tapes to fabricate soft luggage of different shapes and sizes. The fabric cut-pieces may be printed as per customer requirement before stitching.

Economic aspect :

i.	Land and building	: Covered area about 50 sq m. at Rs. 2000 per sq. m. costing Rs. 1.0 lakh
ii.	Machinery and equipment	: Rs. 30,000
iii.	Manpower and wages	: Rs. 20,000 per month
iv.	Working capital (for [illegible] months)	: Rs. 2.5 lakh
v.	Cost of production per annum	: Rs. 11.8 lakhs
vi.	Production capacity	: 27,500 units of bags
vii.	Break even point	: 51.95%

Production of Jute-Synthetic blended 'CAPLON' Blanket in Handloom Sector :

There is a large demand for blanket of intermediate price range in India. Commercially, available products in this range are generally manufactured either entirely from shoddy-recycled low quality wool or its blend with synthetic or other fabrics. Thus a buyer of shoddy blanket is unknowingly paying for used material. 'CAPLON' blanket may be manufactured from fresh and indigenous materials i.e. jute and polypropylene and acrylic. The product has the dual advantage in that the hygiene aspects are being taken care of by the jute component in the hand and warmth and comfort by the synthetic fibre on the other. The thermal insulation value and other properties of 'CAPLON' blanket compare favourably with those of synthetic and woolen blankets whose price are much higher than the foreign goods. The process involved is weaving of jute and synthetic yarns by handloom or power loom. The yarns are blended and dyed before weaving.

Economic aspects :

i. Land and building	:	Land area 250 sq m. Costing Rs 25,000 and covered area 1200 sq m. amounting Rs. 4.8 lakh
ii. Machinery and Equipment	:	2.20 lakh
iii. Labour and staff	:	Rs. 15,000 per month
iv. Working capital	:	Working capital (for 3 months) Rs. 8.6 lakh
. Cost of production per annum	:	Rs. 38.47 lakh
vi. Production capacity	:	16,800 units
vii. Break Even point	:	38.04%

Conclusion :

Future of jute depends on better exploitation of intensive qualities of the fibre and application of appropriate technological inputs. Development and acceptance of jute diversified products have opened up new opportunities in rural sector of our country.

CHAPTER 11

PROBLEMS AND PROSPECTS OF JUTE CULTIVATION IN INDIA

Jute and mesta are the two most important commercial crops in India next to cotton. In trade and industry both the crops together is known as raw jute. The fibre is obtined from the stem (bark) as is known as bast fibre or phloem fibre. The jute fibre is obtained from two species belonging to Tiliaceace family viz. *Corchorus capsularis* known as white jute and *Corchorus olitorius* known as tossa jute. White jute is obtained from two species belonging to Malvaceae family viz. *Hibiscus canabinus* and *Hibiscus subdariffa* var. altissima known as mesta or bimli in India and as kanaf in other producing countries. However, *H. subdariffa* is also known as resella.

The share of jute in total raw jute production is about 85 percent while that of mesta is about 15 percent. Jute and mesta area shares an insignificant area to the tune of 0.5 and 0.1 percent, respectively to the total cropped in the country. However, in the jute growing areas the share of the crop is 1.4 percent to the total cropped area. With its such insignificant coverage, it plays a prominent role in the country's economy by generating employment, earning foreign exchange etc.

Jute Growing Areas and Trend in Area and Production :

The share of coverage under jute and mesta as compared to gross cropped area in the country is only about 0.64 percent. The cultivation is spread in the south Gangetic basin from north-eastern fringe of Uttar Pradesh through Jharkhand to almost all district of West Bengal except the northern hill areas and the western drought prone red and laterite soil areas ; in the Bramha putra basin of Assam extending south upto

Tripura and coastal districts of Orissa . However, the mesta cultivation is spread through out the country. The jute growing states are West Bengal, Assam, Bihar, Orissa, Uttarpradesh, Meghalaya and Tripura. But the major growing state is West Bengal contributing about 68 percent of area and 74 percent of jute production. The major mesta producing state is Andhra Pradesh contributing about 40 percent area and 52 percent mesta production. The statewise normal (9th plan) area and production is given in Table 1.

Presently, the area under jute and mesta has been stabilised at nearly 8 lakh and 2 lakh hectare, respectively and the production of raw jute is around 93 lakh bale. State wise average area and production of jute and mesta since independence have been given in Table 2.

Prospects of Raw Jute :

Jute cultivation in the country started early in the 19th century. Before 18 hundred it was even not known to the people of India. Towards of the end of 18th century the East India Company engaged Mr. Roxburg (A Botanist) to identify some fibre crops which can be used for the manufacture of packaging materials. He identified fibre crops and one of them was installed in 1855 (Auckland Jute Mill). At the beginning of 19th century, the jute occupies the foremost position next to cotton in he industrial sector and also in national economy. Though jute has faced a severe competition during the last two decades, but it has now reviving its importance and has got a better prospect in National and International prospective.

Jute as a natural fibre has definite inheritage advantage. Its silky lusture, high tensile strength, low extensively considerable heat and fibre resistance and long staple length are qualities that makes it suitable for industrial use. Jute is mainly used in he industry in the manufacture of packaging materials. The major jute products are jute twine, yarn, hessian, sacking, carpet backing, clothes etc. These items catered to the packaging requirements of the sectors like food grain, sugar, cement, fertilizer, cotton packaging, salt, postal canvass, vegetable etc.

Diversified Use :

In view of the competition of the jute sector in manufacturing traditional products and also the suitability of jute fibre for the manufacture of various other utility products, initiatives have been taken to use raw jute for manufacture of various diversified products,. These are (i) Jute–geo-textiles; (ii) All jute pile carpets ; (iii) Paper and paper pulp; (iv) Apparel cloth either from pure jute or in blend with other textile fibres; blanket, bed linen made of jute yarn blended with wool or synthetic fibres ; (vi) Coloured /printed shopping bags; (vii) Rigid packaging of teas and apples ; (viii) Soft luggages ; (ix) Handicraft ; (x) Shoe and shoe upper ; (xi) Non-fibrous products including automobile panels ; (xii) Jute composits ; (xiii) Particle board from jute stick for partition wall, table tops, false ceiling etc.

Foreign Exchange Earning :

Near about 15 per cent of the jute products are exported in about 120 countries of the world. The average national exports during the five years (1992-97) is near about 500 crores. The major exporting commodity are carpet backing, hessian and most of the diversified products.

Socio-economic Importance :

i) **Employment generation :** Jute being a labour intensive crop, cultivation of jute creates huge employment opportunities in rural areas. It has been estimated that jute and mesta cultivation can generate about 25 crore working man-days annually. About 40 lakh farm families derive their substance by cultivating jute and mesta in the country. This apart about 2.5 lakh people in the jute industry and 25 lakh people are engaged in jute based ancillary sector.

ii) **Source of fuel :** In the jute growing rural areas, jute sticks are the main source of fuel. In view of the shortage of wood and coal, jute cultivation occupies an important position as a source of fuel in rural areas.

iii) **Soil fertility and crop rotation :** After harvesting the jute crops, the leaves are allowed to shed in fields which is decomposed in the soil and served as a source of manure resulting in an increase in soil fertility. In the jute field the jute follows rice in many of the areas. At the time of harvesting the jute fields remain almost free of weeds and resulting less weed competition in the following crops.

iv) **Environment friendliness :** In view of the recent concern of the people for environment jute has gained lot of importance in this regard. Jute is a natural and renewable source of raw materials and bio-degradable. Considering the above aspects jute is considered an environment friendly crop.

Production System and Problems in Jute Production :

Both the two species of jute are grown in a particular set of edapho-climatic condition utilizing the natural and resources particularly in the eastern and north-eastern part of the country. The crops are grown in different systems and also faced various problems in cultivation which may be summarized below.

i) **Environmental factor :** Jute is cultivated to a particular edapho-climatic condition while the mesta crop is raised in a wide range of agro-climatic conditions. Warm and humid climate with alternate rain and sunshine are most suitable for jute crop. Mesta is a harder crop and can grow in dried condition also. The sowing of jute commended in end of February /early March and continued upto end May and even sometime upto first week of June. Only about 15 to 20 percent of the jute crop is grown under irrigated condition and rest of the area is rainfed . During the sowing period of jute the weather is erratic and crop faced a lot of problems due to weather obligation. The crop can be grown in low medium and high land situations, both moisture stress and stagnant water condition. White jute adopted well under low land situation. While tossa jute is grown in mid and high land situation and so is mesta. About 20 to 25 percent jute is grown under low land situation.

ii) **Crop competition :** Jute crop very often has to face competition with other crops depending upon various factors particularly price of crop and weather factor. In low lying areas the crop has to face competition with deep water rice. In mid and high land jute has to face competition with summer rice and pre-kharif rice, summer vegetables particularly in Assam, West Bengal and Bihar. Mesta is mainly gown in poor land and the competition is less. However, it has to face strong competition with groundnut in Andhra Pradesh and Orissa. The competition with different crops depends mainly on the profitability of the crops.

iii) **Social factor :** Jute is predominantly small and marginal farmers' crop and they raised the crop mainly depending on family labour. About 56 percent area under jute is covered by farmers having less 2 ha holding size. Though sometime jute does not give economic return but the farmer raised the crop to utilize their family labour and also to get the jute stick as a source of fuel.

Use of Seed and Other Inputs :

Though jute is grown mainly in the eastern and north-eastern states but the major seed producing areas are Maharashtra and Andhra Pradesh. As a result the availability of seeds are sometimes delayed. Sufficient quantity of certified seed is not available. About 40 to 50 percent area is covered with certified jute seed. Utilization of fertilizer and plant protection measure are low for the crop. Farmers are very often reluctant to use the various recommended doses of fertilizer because of uncertainty in the prices of their produce.

Technology Adoption :

The technology for cultivation of jute has been developed by the farmers in complete form. Only 5 to 10 percent of the total jute area is line sown. Due to shortage of sowing span, small size of the seed, non-availability of the implements are the main hinderan in line sowing of jute crop. New varieties should be developed depending upon the location specific situation, uncertainty in the return is also a factor.

i) **Retting :** Retting is the important criteria for producing good quality fibre. For retting, farmers generally depend on local natural facilities which are not always acceptable for good quality fibre. Hardly 5 to 10 percent of the crop is retted in canal and river water but most the jute are retted in the water of ponds and depressions. Retting becomes acute in low rainfall season.

ii) **Marketing :** About 95 percent of the raw jute produced by farmers are sold out and only about 5 percent is retained by them for domestic consumption. Farmers do not have the holding capacity of the fibre and are forced to dispose off immediately after the harvest of the crop.

TABLE 1. Statewise normal (9th Five year Plan average) area, production and yield of jute and mesta (Area in '000 ha, production in '000 bale and yield in kg ha^{-1})

States	Jute				
	Area		Production		Yield
	Actual	% share of All India	Actual	% share of All India	
1. Andhra Pradesh	–	–	–	–	
2. Assam	77.0	8.9	720.0	7.5	1683
3. Bihar	144.2	16.8	1041.8	10.9	1301
4. Meghalaya	4.2	0.5	30.9	0.3	1324
5. Orissa	6.9	0.8	56.5	0.6	1474
6. Tripura	1.3	0.2	11.0	0.1	1523
7. Uttar Pradesh	0.04	Neg.	0.3	Neg.	1350
8. West Bengal	626.5	72.8	7756.3	80.6	2228
9. Others	0.5	Neg.	1.8	Neg.	648
10. All India	860.5	100.0	9618.6	100.0	2012

Contd......

States	Mesta				
	Area		Production		Yield
	Actual	% share of All India	Actual	% share of All India	
1. Andhra Pradesh	77.7	41.6	616.2	56.1	1427
2. Assam	5.2	2.8	25.9	2.4	897
3. Bihar	22.6	12.1	156.1	14.2	1243
4. Karnataka	4.5	2.4	6.5	0.6	260
5. Madhya Pradesh	1.9	1.0	5.3	0.5	502
6. Maharastra	28.3	15.2	41.9	3.8	267
7. Meghalaya	4.6	2.5	21.8	2.0	853
8. Orissa	28.6	15.3	118.5	10.7	746
9. Tripura	1.9	1.0	14.8	1.3	1402
10. West Bengal	10.0	4.5	85.3	6.4	1536
11. Others	1.3	22.6	6.1	6.0	877
12. All India	186.5	100.0	1098.4	100.0	1060

TABLE 2. Trend in area, production and yield of Jute and Mesta in India

Year/Plan	Area ('000 ha)		Production ('000 bales)		
	Jute	Mesta	Jute	Mesta	Total
1947-48	263.9	–	1671.3	–	1671.3
Pre-plan (1947-51)	411.3	–	25450	–	2545.0
First plan (1951-56)	645.4	158.3	3928.8	682.6	4611.4
Second plan (1956-61)	704.2	308.2	4441.0	1409.4	5850.4
Third plan (1961-66)	847.2	380.9	5683.6	1648.9	7332.5

Plan holiday (1966-69)	734.5	307.4	4869.8	1133.0	6002.8
Fourth plan (1969-74)	765.0	322.1	5495.1	1220.7	6715.8
Fifth plan (1974-78)	695.9	341.4	4906.2	1593.7	6715.9
Annual plan (1978-80)	859.4	381.5	6270.9	1876.4	8147.3
Sixth plan (1980-85)	818.5	311.7	6419.5	1422.9	7842.4
Seventh plan (1985-90)	802.8	266.1	7562.6	1277.1	8833.7
Annual plan (1990-92)	826.5	238.6	8426.5	1330.1	9756.6
Eighth plan (1992-97)	766.6	196.9	8169.4	1116.7	9286.1
Ninth plan (1997-02)	860.6	186.6	9618.6	1098.4	10717.0

Year/Plan	Yield (kg ha^{-1})	
	Jute	Mesta
1947-48	1140	-
Pre-plan (1947-51)	1114	-
First plan (1951-56)	1096	776
Second plan (1956-61)	1135	823
Third plan (1961-66)	1208	779
Plan holiday (1966-69)	1193	664
Fourth plan (1969-74)	1293	682
Fifth plan (1974-78)	1269	640
Annual plan (1978-80)	1314	885
Sixth plan (1980-85)	1411	822
Seventh plan (1985-90)	1696	864
Annual plan (1990-92)	1833	1004
Eighth plan (1992-97)	1918	1021
Ninth plan (1997-02)	2012	1060

TABLE 3. Countrywise normal (Average of 1997-2002), Area, Production and Yield of Jute and Jute like fibre (Mesta) in the world

Country	Area ('000 ha)		Production ('000 bales)		Yield (kg ha^{-1})
	Actual	% to the world	Actual	% to the world	
1. Bangladesh	474.9	26.7	4740.8	26.2	1769
2. China	84.1	4.7	1227.7	6.8	2628
3. India	1047.0	58.8	10717.0	59.3	1842
4. Myanmar	39.1	2.2	201.5	1.1	928
5. Nepal	12.2	0.7	85.4	0.5	1260
6. Thailand	30.9	1.7	282.0	1.6	1643
7. Others	91.8	5.2	809.6	4.5	1587
8. World total	1780.0	100.0	18064.0	100.0	1827

❑❑❑❑❑
❑❑❑